ENEMIES OF PATIENTS

ENEMIES OF PATIENTS

Ruth Macklin

New York   Oxford
OXFORD UNIVERSITY PRESS
1993

Oxford University Press

Oxford   New York   Toronto
Delhi   Bombay   Calcutta   Madras   Karachi
Kuala Lumpur   Singapore   Hong Kong   Tokyo
Nairobi   Dar es Salaam   Cape Town
Melbourne   Auckland

and associated companies in
Berlin   Ibadan

Copyright © 1993 by Ruth Macklin

Published by Oxford University Press, Inc.,
200 Madison Avenue, New York, New York 10016

Oxford is a registered trademark of Oxford University Press

Library of Congress Cataloging-in-Publication Data
Macklin, Ruth, 1938-
Enemies of patients / Ruth Macklin.
p. cm.   Includes bibliographical references and index.
ISBN 0-19-507200-6
1. Medical ethics.   2. Physician and patient.
I. Title.   R724.M16135 1993
174'.2—dc20 92-16894

1 3 5 7 9 8 6 4 2

Printed in the United States of America
on acid-free paper

For Meryl and Shelley

Preface

As a philosopher working in the field of biomedical ethics for more than 20 years, I have become concerned by a number of developments that are increasingly hostile to the interests of patients. Some of these developments are the result of wrongheaded responses to concerns about money: unwarranted fears of liability on the part of hospital administrators and physicians' growing belief that they are obligated to stem the tide of rising medical costs by rationing care at the bedside. A different, perhaps ironic development is the emergence of laws or regulations designed to protect patients but yielding unanticipated, burdensome consequences. Still other developments can be traced to physicians' reactions to a new disease—AIDS—and to problems surrounding allocation of new medical technologies.

These forces intrude into the physician–patient relationship, restricting the right of patients to exercise self-determination. Beyond that, these intrusions sometimes limit the ability of doctors to act in the best interest of their patients, and more frequently cause doctors to believe that their power and authority are limited. An array of "enemies" of patients is identified by tracing a number of developments that are changing the way medicine has traditionally been practiced.

I have sought to describe the ways in which administrators are replacing physicians as decision makers in a wide array of matters surrounding patient care. One chapter (3) offers a critical analysis of the newly created occupation of "risk manager" in hospitals. This position has assumed increasing importance, eroding the autonomy of both patients and physicians as decision makers. Risk management is now central to the way hospitals are administered. Other chapters analyze the effects in the clinical setting of recent regulations and legislation. Another chapter (4) is devoted entirely to the impact that political and ideological concerns about the fetus have had on the rights of pregnant women, especially in limiting their autonomy.

In the face of all these changes, many doctors continue to be strong advocates for their patients. Newly created entities, such as hospital ethics committees, can also be a positive force for promoting the rights and interests of patients. I have tried to balance the picture of an array of enemies threatening patients by portraying a counterforce of sincere and devoted clinicians, lawyers, and ethicists seeking to do the right thing.

Many of the cases used to illustrate the points in these chapters come from my own experience in clinical ethics. The names of both patients and physicians, as well as some identifying details, have been altered to protect confidentiality. But the events are real and the ethical dilemmas are depicted in precisely the manner in which they were presented by physicians and medical students at case conferences, ethics consultations, and meetings of ethics committees. Additional cases and issues are drawn from the scholarly and professional literature, and readers will probably recall some stories from accounts in the media.

My educational background is in philosophy. Therefore, in writing this book, I have not used the methodology of the social sciences: data gathering, ethnology, and quantitative or qualitative sociological research. Instead, I have used a well-known method of philosophy: case studies that serve as paradigms, followed by an ethical analysis. I ask readers not to view the cases as mere anecdotes or misconstrue them as anecdotal evidence in the absence of scientifically gathered data.

The method of paradigms uses clear cases to exemplify situations recognizable to clinicians and to other bioethicists who work in hospitals. Paradigms do not work as substitutes for data that might be gathered about the practices or policies described. They are not the equivalent of statistical generalizations about any of the so-called enemies of patients—how widespread they are or how prevalent their intrusions may be. Instead, presenting clear cases makes it possible to convey a richness of detail and a sense of the reality patients and clinicians experience. Paradigms are the best vehicle for providing an ethical analysis, drawing on leading principles of bioethics and other prominent social values. Finally, the method of paradigms invites consideration of the less clear examples and cases having shades of gray, where ethical uncertainty prevails.

The material in this book should be of interest to readers from many walks of life.

New York   R.M.
April 1992

ACKNOWLEDGMENTS

I am grateful to my colleagues at Albert Einstein College of Medicine and to the medical students and postgraduate trainees for their thoughtful and open discussions of the cases presented in this book. My thanks also to the patients, whose characteristics I have altered slightly in order to protect their confidentiality.

Some material in this book is adapted or excerpted from previously published articles. Portions of Chapter 4 appeared in my article "Maternal-Fetal Conflict: An Ethical Analysis," *Women's Health Issues*, vol. 1, no. 1 (1990). A brief section of Chapter 7 is taken from "Is It Ethical to Ration at the Bedside?" *Medical Ethics for the Pediatrician*, Vol. 6 (April 1991). Sections of Chapter 9 are published in my chapter, "Which Way Down the Slippery Slope? Nazi Medical Killing and Euthanasia Today," Arthur L. Caplan (ed.), *When Medicine Went Mad* (Totowa, NJ: Humana Press, 1992). And portions of Chapter 10 are excerpted from "The Inner Workings of an Ethics Committee: Latest Battle over Jehovah's Witnesses," *Hastings Center Report*, vol. 18 (February/March 1988). I thank the editors and publishers for permission to use the material.

My thanks also to medical colleagues who made helpful comments and criticisms of drafts of several chapters: Alan Fleischman, M.D., Saul Moroff, M.D., Steven Martin, M.D., and Matthew Berger, M.D. Special thanks go to my friend, Dr. Marie Burnett, for her thorough editorial work on the entire manuscript and for her moral support. The encouragement and insights of Jeffrey W. House at Oxford University Press were a great asset from beginning to end, and I am grateful to Helen Greenberg for her skillful editorial work on the final version of the manuscript.

The individuals depicted in this book as "enemies" of patients will probably not want my thanks. Nevertheless, I acknowledge their role in prompting me to think about enemies and advocates of patients.

Contents

ENEMIES OF PATIENTS

1

Enemies and Advocates of Patients

During a heated discussion at a meeting of a hospital ethics committee, a nurse proclaimed that she and members of her profession serve as "advocates" for patients. An irritated physician, speaking for his profession, replied: "Then who are we? Enemies of patients?"

Of course, that doctor meant this to be a joke. But there was a bitter edge to his comment, which revealed the frustration he and other physicians feel when patients lack gratitude, question their doctors' recommendations—sometimes even refusing treatment—and sue the doctor if things fail to turn out perfectly. That doctor was justified in defending the medical tradition in which physicians have long been trained to advocate fiercely in their patients' interests.

It is probably an irony of the recent period, in which patients have been given a greater voice in decision making, that physicians have come to be viewed as adversaries. Nurses, whose role has been that of caring, in contrast to the physicians' role of curing, have come to see themselves as advocates of patients' rights and interests, as have hospital social workers and individuals specifically trained or designated as patient advocates. The reason for the shift is the perception that physicians, although still committed to serving their patients, are advocating for patients' best *medical* interests, while the other groups advocate for patients' *rights*, especially their right to participate fully in decision making related to their own care and treatment.

Disagreements alone do not make people into enemies. Even when physicians and patients disagreed about what course of treatment to pursue, for the most part doctors were perceived as acting in what they sincerely believed was their patients' best interests. The traditional view holds that patients' interests lie in continued medical treatment and in life-prolonging therapy. In promoting those interests, physicians adhered to a noble ethical principle, the principle of "beneficence," which directs doctors to do good. They gave this principle priority over another sound but sometimes competing principle, "respect for autonomy," which recognizes the right of patients to be full participants in decisions regarding their treatment.

Doctors are supposed to act in the best interest of their patients. Perhaps

3

that is an idealized image of the physician, but it does not seem controversial to assert that physicians should strive to reach that ideal even if it cannot be fully attained. The avowed ethic of medicine has always been patient-centered. The physician–patient relationship is the type known as "fiduciary," characterized by trust on the part of patients or clients and a duty on the part of professionals to act in the patient's or client's interest rather than in their own self-interest.

As with most ethical ideals, real-world circumstances often make it difficult or impossible for physicians to serve their patients' interests with an unswerving commitment. These circumstances may stem from a clash between a doctor's personal values and those of the patient, making it unclear how to define the patient's "best interest." Or the circumstances may arise out of institutional barriers to physicians' ability to advocate for their patients. Even broader circumstances stem from societal changes in the way medicine is practiced, forcing doctors to subordinate the interests of their patients to other pressing concerns.

Looking at the physician–patient relationship from an ethical point of view, it is important to distinguish situations in which doctors act in ways they honestly believe are required by their professional obligation to patients from those in which their ability to fulfill their obligation to patients is compromised by outside forces. The latter circumstances constrict the professional autonomy of physicians and lessen their ability to advocate fully for their patients. But the first type of situation still creates tensions and deserves ongoing examination. It is exemplified in the case of Mrs. S.

Mrs. S was a 97-year-old woman who had been living in a residence for seniors. Up to three years ago, at age 94, she had been living independently, still riding the New York City subway. Two years ago she suffered a stroke. She developed difficulty in swallowing, and her nutritional status became compromised. She had never before been hospitalized. Now she was admitted to the acute care hospital with pneumonia, fever, and difficulty in swallowing. She was not responding to antibiotics. Since her poor nutrition contributed to some of her medical problems, the house staff—residents and interns responsible for her hospital care—recommended that she be given artificial nutrition through a nasogastric tube, inserted into her nose and down to her stomach, or else through a catheter inserted into a large vein in her neck.

At this point, communicating with Mrs. S had become a problem. Her two daughters, who were involved in the decision-making process, resisted the doctors' plan to begin invasive procedures to feed their mother. They reported that she had signed a living will, stating that she did not wish to have any "heroic measures or invasive procedures" to prolong her life if she developed a terminal illness. Mrs. S's daughters knew their mother's wishes, since she had presented her living will at a family gathering 11 years ago.

One daughter pointed out that over a period of 50 years. Mrs. S had been a secondary school teacher and then a principal. An independent and decisive person and a world traveler, she had remained mentally alert up to very recently, when she became deaf and had trouble communicating. She had trusted her daughters to be her advocate in case she became unable to make her own medical decisions. Her personal physician was also aware of her wishes about invasive, life-prolonging measures and had written a note in the hospital chart. One of her daughters was present at a medical conference devoted to her mother's case.

At the conference, one doctor spoke about the effect that cases like this have on medical personnel. He said that the interns and residents—young physicians in training—would suffer the psychological burden of being responsible for the patient's death if they failed to begin artificial feeding. People must be aware, he said, of the physician's emotional response to caring for terminally ill, elderly patients. The interns and residents insisted that pneumonia is a treatable illness, and therefore it was their obligation to treat this patient. They were relying on the notion that "doctors always have to treat what's treatable." Beyond that, they had aroused feelings of guilt in the patient's daughters for refusing to allow the artificial feeding they proposed. The daughter present at the conference said she had been made to feel that she was advocating "killing her mother."

Should the emotional tension felt by the interns and residents be factored into the decision-making process? Is it even ethically relevant? Doctors sometimes argue that it would be morally wrong for them to violate the dictates of their conscience, that they are not mere technicians, bound to obey blindly the wishes of patients or their families. Although that is surely true, the ethics of medicine must remain patient-centered. Unlike situations where the individuals involved are roughly equal in power or vulnerability, the physician–patient relationship is not one in which the feelings and values of both parties have somehow to be balanced.

And how about Mrs. S's daughters, who appeared to be loving and supportive? Do the requests of family members count as much as a patient's expressed wishes about forgoing treatment? The circumstances of Mrs. S's case gave no reason to question her daughters' motives. They were seeking to act as surrogates for their mother, yet they were being thwarted by zealous young doctors in the hospital who took a different view of the nature of their ethical obligations.

Is there a single right answer to the dilemma posed by the case of Mrs. S? The situation seems ethically ambiguous. The patient had made a living will, which provided evidence of her wishes. She made those wishes known to her daughters and placed her trust in them to help carry out her desires. The ethical principle that governs this situation is *respect for persons*, mandating recognition of a patient's autonomy. Translated into law, this ethical principle grants people the right to self-determination in the medical setting.

But how clearly did the living will state Mrs. S's wishes? The words used to describe the treatments she did not want were "no heroic measures or invasive procedures." This points to a classic problem that has plagued the traditional living will. Is a feeding tube a heroic measure or an invasive procedure? Is that what Mrs. S had in mind when she made out her living will 11 years earlier? Since her living will stipulated that heroic measures and invasive procedures were not to be used if she became terminally ill, was it accurate to describe her condition now as a terminal illness? These are some of the uncertainties in living wills that fail to specify the particular treatments a patient would or would not want. An additional uncertainty arises when the proposed heroic or invasive treatment may be only temporary. If the treatment results in the patient's improvement, a good quality of life might be restored.

This possibility points to another reason why the case of Mrs. S appears ethically ambiguous: respect for autonomy is not the only moral principle of biomedical ethics. The principle of beneficence, which directs physicians to maximize benefits and minimize harm to the patient, can conflict with the autonomy principle. The interns and residents who insisted on treating Mrs. S felt that their obligation derived from the principle of beneficence.

Although prolonging life is normally a benefit to the patient, there may come a point where continued life is more of a burden than a benefit. In Mrs. S's case, an experienced physician observed that even with artificial feeding, the chance of her dying was about 90 percent. And surely the patient's quality of life would continue to diminish once the tubes or catheters were inserted. So the ethical principle that mandates maximizing benefits to the patient and minimizing harm would not be violated if tube feeding were withheld, since it would not truly benefit Mrs. S to have her life prolonged by this means.

The case of Mrs. S appeared at first to be ethically ambiguous, but in the final analysis the resolution is clear. The two ethical principles—autonomy and beneficence—are not in conflict but are mutually reinforcing. In Mrs. S's case, the wishes expressed in her living will, her values expressed over a lifetime, her current deteriorating condition, and the minimal benefits that continued medical treatment could offer all point to the conclusion that withholding artificial feeding is the right thing to do.

Empowering Patients: Two Decades of Progress

The past 25 years have seen striking changes in the way medicine had been practiced for centuries. No longer is the physician a father figure making decisions for patients without their knowledge and permission. No longer may a physician withhold a diagnosis of cancer from a patient without being criticized as paternalistic—presuming to know what is best for the patient. In contrast to medical practice as recently as the 1950s, it is now illegal, as well as unethical, to conduct medical research without the informed consent of human subjects. Informed consent from patients is also required for a number of procedures in addition to surgery.

Hippocrates, the Greek physician sometimes called the father of medical ethics, admonished physicians to perform their duties

> calmly and adroitly, concealing most things from the patient while you are attending to him. Give necessary orders with cheerfulness and sincerity, turning his attention away from what is being done to him; sometimes reprove sharply and emphatically, and sometimes comfort with solicitude and attention, revealing nothing of the patient's future or present condition. [1]

With some modifications, this Hippocratic tradition has reigned for over 2,000 years. A study published in 1961 showed that slightly over 88 percent of physicians at that time withheld from their patients the fact that they had cancer. And as recently as 1971, an editorial in the *Journal of the American Medical Association* questioned the obligation of doctors to tell a patient with a fatal disease the truth. The editorial was entitled "Must the Physician Set Himself Above Nature and play Supergod?"

Now, however, much has changed. There is a widely recognized *right* of patients to know their diagnosis and prognosis. When the 1961 study of what physicians tell cancer patients was redone in 1977, 98 percent reported that their general policy is to tell the patient. This does not mean that physicians have an obligation to inflict unwanted information on their patients. But it does mean that the patient—not the doctor—should be the one to judge when and how much information should be disclosed.

Even when it comes to life-sustaining treatments, most doctors have ceased to believe that they have a moral obligation to prolong a patient's life despite the patient's wishes. Both experienced physicians and recent graduates of medical school have, for the most part, come to respect their patients' right to participate in decisions surrounding their care, including the right to refuse life-prolonging treatment. Laws in most states recognize the right of adult patients who have the capacity to make decisions to refuse medical treatments. The concept of a "living will" has become well established, granting people the right to specify what treatments they would or would not want, or whom they want to make medical decisions on their behalf once they become incapacitated. As of April 1992, all 50 states had enacted some form of legislation recognizing advance directives in the form of living wills or appointment of a health care agent or proxy for future decision making.

Today, patients have rights. Yet there are some who question whether the pendulum has swung too far in the direction of patients' rights, and whether it is now time to limit the authority of patients to direct their own care in an almost limitless way. Not infrequently, physicians are unwilling to honor patients' rights. One doctor stated at an ethics committee meeting that some years ago, his disclosure of the diagnosis of cancer led his patient to commit suicide. At the time the diagnosis of metastatic liver cancer was made, nothing more could be done for the patient. Based on that episode, this physician reported that he is always reluctant to disclose a diagnosis of cancer to his patients. "I still have trouble sleeping at night," he said.

What conclusion should be drawn from this case? One possible conclusion is that the physician was mistaken in his assessment of what was in his patient's best interest. Is suicide always an act to be prevented or deplored? The patient with incurable, advanced liver cancer who decided to take his own life chose the time and manner of his death. Even if the doctor who made the disclosure continues to feel guilty, it does not follow that informing the patient of his diagnosis was a mistake to be regretted.

A second conclusion is that the case was only one incident in one physician's experience. Although his memory of that experience remained vivid, it is best viewed as an anecdote from which it would be a mistake to generalize. The plural of "anecdote" is not "data." More objective information was gained in a survey conducted by the President's Commission for the Study of Ethical Problems in Medicine and Biomedical and Behavioral Research. That study provided little evidence to support the view held by doctors that disclosing a terminal illness to patients has as dire consequences as doctors often assume. The President's Commission's report stated:

> Despite all the anecdotes about patients who committed suicide, suffered heart attacks, or plunged into prolonged depression upon being told "bad news," little documentation exists for claims that informing patients is more dangerous to their health than not informing them, particularly when informing is done in a sensitive and tactful fashion. [2]

Isn't it true, nevertheless, that informing a patient of a diagnosis of a fatal illness usually causes the patient great emotional pain? Who among us can receive bad news about our health or prognosis without suffering distress? A classic response to the insistence that patients have a right to know information about their medical condition is the question "What about the patient's right *not* to know?"

Physicians have an obligation to be truthful to those patients who indicate that they do want information, but there is surely no obligation to inflict unwanted information on patients who give signs that they do not wish to hear bad news. Furthermore, the way the information is imparted can make all the difference. "Truth dumping" has a deservedly bad reputation because it demonstrates insensitivity. In contrast, "titrating" the dose of unpleasant news, like titrating potentially toxic doses of medication, provides the information to patients a little at a time, over time, and can succeed in telling them all they need to know and all they wish to know in the way that causes least harm.

Patient's rights have been granted a semiofficial status in various pronouncements. In a statement first issued in 1973, the American Hospital Association published "A Patient's Bill of Rights."[3] The statement is noteworthy not only for the broad array of rights it appears to grant to hospitalized patients, but also for the gap between these ideal rights and the real world of many hospitals. The first right affirmed is: "The patient has the right to considerate and respectful care." Anyone who has been a patient in a hospital might reasonably wonder whether this and other rights in this Bill of Rights are taken seriously by hospital staff. Yet as a statement of ethical ideals, A Patient's Bill of Rights gives health care professionals something to strive for. A plaque enunciating the 12 statements is posted on the walls of most hospitals and provides valuable information to patients who may wish to consult someone in the Patient Representative's office.

Included among the other rights enunciated in this document are the right to obtain from one's physician "complete current information concerning . . . diagnosis, treatment, and prognosis in terms the patient can be reasonably expected to understand"; the right to information necessary to give informed consent prior to the start of any procedure; the right to refuse treatment to the extent permitted by law; the right to privacy and to confidentiality of "all communications and records pertaining to his care"; and the right to expect reasonable continuity of care.

Besides the gulf between the rights spelled out in these statements and much of what actually goes on in hospitals, there are other gaps between ethical ideals and the realities of medical practice. It is still true that many physicians are skeptical about the meaningfulness of obtaining a patient's informed consent to treatment. They contend that patients are sometimes too sick, too ignorant, or too frightened to understand what they have been told in the process of providing information. Informed consent to treatment remains a largely unrealized ideal.

In addition, despite a great deal of publicity about the desirability of making advance directives—writing a living will or appointing a surrogate to carry out one's wishes about medical treatment in the event that one becomes incapacitated—too few people have taken those steps to make it likely that their wishes will be honored. Generally speaking, doctors are of little help in this regard. Many physicians are reluctant to initiate discussions with patients about preparing advance directives, partly because doctors have not been trained to do so and partly because physicians themselves are uncomfortable discussing issues like terminal illness and death. Furthermore, doctors worry that patients will assume that the reason for bringing the subject up is that there is some bad news about their medical condition or prognosis that the doctor has been concealing.

Despite these and other gaps between moral ideals and actual practice, the changes brought about in the past 20 years have been salutary. In medicine as elsewhere, moral progress cannot occur without a conception of an ideal to strive toward. Although that ideal is far from being fully attained in today's health care system, a greater recognition of patients' rights is a measure of considerable moral progress.

Still, there are some who question whether granting decision-making rights to patients represents moral progress. Doubts have been voiced about whether this expansion of patients' rights is a good thing. One physician wrote that the proposition that these newly established patients' rights are good is far from a self-evident truth. He asserted that "many educated people think that rights are not necessarily empowering and that the patient rights movement has actually disempowered patients precisely by undermining the trust developed between physician and patient.[4] These challenges deserve a response. As a philosopher working in the field of bioethics, I hold no propositions to be self-evident except, perhaps, the truths of mathematics and deductive logic. All ethical propositions require some justification. However, the deeper such propositions are rooted, as they descend to the very foundation of ethics, the harder they are to justify in terms of more fundamental propositions.

General skepticism about the patients' rights movement does seem puzzling, since it runs counter to a cluster of long-cherished values related to individualism in democratic societies: the right to self-determination, personal autonomy, and individual liberty. The skeptic would have to argue that despite the importance accorded these values in other human endeavors, they are of little import in the physician–patient relationship. A self-serving spokesperson for the medical profession might take that position, as might physicians dismayed by the erosion of their power or authority vis-à-vis their patients. If the skeptic means to go so far as to question why liberty, autonomy, and self-determination should be accorded the importance they have been given in Western moral philosophy, a separate treatise would be required by way of response.

Nevertheless, there is a compelling ethical argument underlying the skeptic's position regarding the place of autonomy and self-determination in the patient–physician relationship. One justification for questioning whether the creation and expansion of patients' rights is a good thing is that it has resulted in conflicts with patients' best interests. Physicians find themselves struggling to establish the proper balance between their obligation to do what they believe to be medically best for patients and their obligation to respect the patient's wishes.

Situations in which patients' rights conflict with their real or apparent best interests tend to turn doctors and patients into adversaries. For the most part, however, doctors and patients are allies in the fight against disease. Physicians make treatment recommendations that patients readily accept after being more or less informed about the risks, benefits, and alternatives.

There may be another legitimate ground for skepticism on the part of physicians about whether establishing patients' right to decision-making autonomy has been a good thing. If it is true, as some have argued, that adopting the language of rights tends to put people in an adversary posture, it could lead to a further erosion of the physician–patient relationship. So, although granting rights to patients can succeed in serving the legitimate interests of particular patients who exercise their rights, the movement may strike a blow at patients in general by undermining the very nature of the doctor–patient relationship. Is this a sufficient reason to lament the patients' rights movement? Perhaps it would be better on the whole for patients—as well as for physicians—to halt the proliferation of rights and seek a return to a less adversarial, if more paternalistic, practice of medicine.

One problem lies in the fact that doctors and patients can and often do disagree about what really is in the patient's interest. This problem is further complicated when patients' rights are brought into the picture. A long-standing example of the conflict between patients' rights and what physicians take to be their best interests is the ongoing problem of Jehovah's Witnesses who refuse blood transfusions on religious grounds. Most of the time, these are otherwise healthy patients who require a transfusion in the course of surgery and could normally be restored to full health. Witnesses consent to the surgery but refuse blood and blood products that may be deemed necessary to preserve their life if they lose too much blood during the surgical procedure.

Surgeons and anesthesiologists are the most vociferous in arguing that this legally established right of Jehovah's Witnesses to refuse a simple, relatively low-risk procedure is in direct conflict with their best interest. But they are not the only ones who make this argument. The patients are individuals who do not wish to die, who come to doctors and hospitals seeking medical treatment, and who are judged to be irrational in their refusal of blood and in placing their religious beliefs ahead of their interest in self-preservation. A few surgeons can be found who are willing to risk the surgery without harboring a covert intention to give blood if it becomes necessary.

Many anesthesiologists are unwilling to participate in cases where patients might die from loss of blood during the procedure. At an ethics committee meeting, one anesthesiologist proclaimed, "I'm not here to see a patient die. I didn't go into medicine for that." When asked what he would do if a Jehovah's Witness categorically refused blood in advance of the operation, the doctor said he would promise to respect the patient's wishes. "But then, if I was wrong in my prediction that transfusion wouldn't be necessary, it would then be an emergency situation and I'd have to give blood." Did he not promise the patient he wouldn't transfuse? "Well, yes, but I honestly believed blood wouldn't be necessary. In the rare circumstance when the prediction turns out wrong, I'd have to transfuse. You don't need to get informed consent in an emergency," the physician said.

It appears that this physician regularly makes a lying promise to his patients, telling them he won't transfuse. Once they are anesthetized, if a loss of a volume of blood creates an emergency situation, he transfuses. Since it is not necessary to obtain a patient's informed consent in an emergency, the physician tries to justify his action by appealing to the emergency exception. But that justification does not apply. The physician cannot be excused for failure to obtain proper consent for a procedure when an emergency could have been foreseen.

A recent experience demonstrates the extent to which some physicians are unwilling to recognize patients' rights that are otherwise widely acknowledged by law, as well as in medical ethics. In a discussion with a group of anesthesiologists about Jehovah's Witnesses and their right to refuse blood transfusions, I mentioned the alternative of honoring the patient's refusal. One physician shouted: "You call that honor? It's murder!" There was no discussion, no willingness to consider whether a patient's informed refusal of treatment could be respected. It was "murder" for any physician to comply. Another anesthesiologist at the conference ventured the thought that it wouldn't really be murder; suicide is a more accurate description. "No," the first physician insisted, "any physician who stands by and lets that happen is guilty of murder!"

Another participant suggested that "if Jehovah's Witnesses want to live in our society, they have to abide by our rules. Otherwise, they should build their own hospitals and supply their own doctors." When I replied by referring to Jehovah's Witnesses as a religious minority, and spoke of the need for the majority in a pluralistic society to respect the rights of minorities, a large chorus of groans was the response. This recent experience with an uncharacteristically rude and hostile group of physicians served to remind me that not all doctors today are prepared to give up their right to practice medicine as they see fit. Claiming that this right is a prerogative of medical professionals, they perceive the situation as a conflict of rights between the doctor and the patient. But there is another, more enlightened way of analyzing the situation.

In one respect, Jehovah's Witness cases do constitute an obvious clash between patients' rights and their own best interests. Assuming, as most people do, that it's in a person's best interest to continue living, the right of Jehovah's Witnesses to refuse medical treatment goes against their objective best interests. In this respect, they might be viewed as patients who become enemies of themselves.

Looked at in another light, however, their rights and best interests are not in conflict. Drawing that conclusion requires that we consider their best interests as Jehovah's Witnesses themselves see them. For them, to accept a blood transfusion is to be cut off from the possibility of eternal salvation. They value life, but they believe in immortal life as well as life here on earth. In fact, they believe they are just "passing through" in this life, which is limited, after all. According to the belief system of Jehovah's Witnesses, it is clearly in their best interests to shorten their remaining mortal life in order to gain a chance at eternal salvation. For the rest of us who do not share their belief system, their best interest lies in accepting the simple blood transfusion that will prolong their mortal life. However, what it means to respect patients' autonomy is that their values should be recognized as valid for them. Only when patients suffer diminished autonomy does paternalism become justified, and only then is it ethically permissible to seek to override their refusal of lifesaving treatment.

As the ongoing struggle surrounding treatment of Jehovah's Witnesses illustrates, the ways in which physicians sometimes respond to patients' legal and moral rights betray a reluctance or hostility to cede ultimate decision-making authority to the patient. But isn't that reluctance understandable, perhaps even admirable, when the likely outcome is the death of a patient? The medical profession has long embodied the conviction that doctors must not be agents or advocates of death. That value is prominent in the medical profession's condemnation of the notion that doctors could administer lethal injections to prisoners given the death sentence. It also underlies the stalwart opposition of most physicians to active euthanasia, despite the mounting societal debate and a number of recent legislative proposals that have sought to legalize physicians' "aid in dying." Still, there is a difference between asking physicians to act as executioners or "euthanizers" and requiring them to refrain from performing invasive or intrusive medical procedures on unwilling patients.

On the whole, there are few doctors today who do not recognize their patients' right to participate in medical decisions surrounding their care. In contrast to the early days of biomedical ethics, when some physicians were puzzled about what so-called ethicists were doing in their midst, today it is not uncommon for educational programs to include a presentation devoted to ethical practice. There are still some physicians who have not emerged from the dark ages of medical practice, in which doctors wielded virtually ultimate power and authority over their patients, as illustrated by the behavior of the rude (and legally uninformed) group of anesthesiologists.

The doctors at this conference held the firm conviction that it is their right to treat patients in a way they deem medically appropriate. Patients who are unwilling to accept their authority can go elsewhere. These anesthesiologists adhered to a long-standing precept embodied in the American Medical Association's statement of ethical principles: "The physician may choose whom to serve." May physicians simply refuse to care for patients? If so, when can their refusal be justified?

Physicians refuse to treat patients for a variety of reasons, including fear of legal liability. A few years ago, newspapers reported that all obstetricians in a large region of Georgia came to an agreement that they would not provide obstetrical services to women who were lawyers, married to lawyers, or worked in any capacity for a law firm. This was discrimination against an entire class of women, stemming from the belief that the likelihood of having to face a lawsuit would drop dramatically if they refused to take as patients any women who were connected in any way with lawyers or law firms. There are several ethically acceptable things doctors can do to lessen the probability of being sued, but refusing their services to an entire category of women is not one of them. It would be incorrect to consider this group of obstetricians enemies of their patients, since they declined even to accept as patients women whom they perceived to be a threat to their interests. They could nevertheless be criticized with unjust discrimination in choosing whom they would serve.

Physicians committed to beneficence can be considered temporary adversaries of their patients, but they are surely not enemies. Physicians become true enemies of patients only when they discriminate against particular patients based on their presumed social worth or when the physician acts in a malevolent fashion. Few doctors are malevolent actors regarding their patients, yet there are exceptions. An example was recounted in my book *Mortal Choices*.

Lori was a 19-year-old single woman hospitalized for orthopedic surgery on her legs and feet to repair a birth defect. She then needed plastic surgery and a skin graft, which were successful. She was about to be discharged but was awaiting a brace and special shoes. Lori had had emotional problems, and her intelligence was judged to be borderline. In the past, she had not complied with her medical regimen, and the social worker recommended home care. Lori needed a home aide or visiting nurse. The social worker obtained the proper papers to be filled out before home care could be provided. The attending physician in the hospital refused to fill out the papers. When the social worker asked why, the physician replied: "It is an abuse of Medicaid." He said he was opposed to having "our tax dollars pay for this."[5] This doctor viewed himself as having an obligation to society, not to his patient. He was wrong medically and ethically.

Intrusions and Impositions by Outside Forces

The struggle to empower patients has not been easy, and it is by no means fully accomplished. Now, however, signs have begun to appear that new and different forces threaten to erode the recently recognized rights of patients. These forces are impinging on the doctor–patient relationship even when that relationship is without conflict, diminishing the authority and traditional professional autonomy of doctors to act on behalf of their patients.

Administrators and hospital attorneys regularly intrude in situations where there is no disagreement between physicians and their patients or families. Fearing possible legal repercussions, hospital administrators sometimes seek to prevent competent adult patients from refusing burdensome treatment even when physicians concur with the patient's wish. These administrators operate at several levels. Presidents or executive directors of hospitals, administrators on duty, and a newly created breed known as "risk managers" make decisions that affect the care and treatment of individual patients within hospitals. Risk managers are administrators, usually without any legal training, whose job it is to decrease the liability of hospitals.

A recent article by Marshall Kapp depicts increasingly common situations in which risk management trumps ethics. One situation involves risk managers urging the continuation of aggressive, expensive treatment for critically ill patients, even when the patients themselves and their families wish the treatment to cease and when physicians judge that continued treatment is unlikely to be medically beneficial for the patient. In this situation, the autonomy of patients and families is violated, the patients do not benefit, and resources end up being wasted.[6]

Why would a risk manager see continuation of treatment as being in the hospital's interest? Wouldn't cessation of treatment be more likely to serve the institution's financial interest? The explanation reveals the fears and misperceptions that infuse the way risk managers do their job. They may fear criminal indictment by a prosecutor who might view termination of life supports as murder, in spite of clear legal precedents permitting such requests when made by competent patients. Or they may fear that some relative will appear and sue the hospital for withdrawing treatment, despite the fact that there is no legal or ethical basis on which a relative can override a competent patient's request. Or the risk manager could be worried about bad publicity for the hospital if the story is broadcast on the eleven o'clock news—an unlikely prospect, to be sure.

Another situation involves the use of physical restraints on nursing home residents. Kapp notes that "Overuse of restraints, that is, their invocation in circumstances where less restrictive or intrusive alternatives are available to accomplish the legitimate goal of resident safety, violates the resident's ethical interests in autonomy, beneficence, and nonmaleficence."[7] Not only do risk managers give advice to nursing home administrators that violates the rights and goes against the interests of residents; many also fail to do their job properly. The article points out that far more lawsuits are filed because of injuries sustained by the inappropriate application of physical restraints than on the basis of failure to apply restraints.[8]

Local Administrators as Final Arbiters

Although it would seem obvious that the mission of hospitals and nursing homes is to care for the sick and the frail, administrative concerns have taken on a life of their own, sometimes thwarting and occasionally even contradicting the chief function of caring for patients. The case of Mr. Romero demonstrates how patients can be both harmed and wronged by the actions of administrators, whose concerns are often at odds with the interests of patients.

Angel Romero, a 56-year-old man, was referred directly to the hospital admissions office by his nursing home.[9] The admitting office immediately contacted the administrator on duty regarding the patient. The administrator then called the nursing home and was informed that Mr. Romero needed to be evaluated for hospitalization, as he was not following "home rules." He had been walking out of the nursing home building, sitting in an adjacent park, and drinking beer outside the nursing home. He was warned that he wasn't following the rules of the nursing home, but he had been there for only three days.

The administrator contacted the emergency room and requested a medical/psychiatric evaluation of the patient. Mr. Romero's old chart indicated that he was suffering from Alzheimer-type dementia and had been discharged from the medical service at the acute care hospital three days earlier, after almost one year of continual hospitalization. He had initially been admitted to the medical service because of uncontrolled blood pressure and seriously disorganized behavior secondary to his dementia. After he was stabilized, the reason Mr. Romero remained in the hospital so long was that he had become a "disposition problem": his family refused to take him back; because of his advanced dementia he could not be sent to a shelter; and social workers were unable to arrange another placement.

On this admission to the hospital, an assessment done in the emergency room showed Mr. Romero to be medically stable. However, he was found to have severe memory impairment and poor reality testing, which his dementia would explain. But there was no evidence of dangerousness to himself or others, or of any further abnormalities that warranted hospitalization. The findings from this examination were essentially the same as those just before his discharge from the hospital three days earlier.

The hospital administrator contacted the psychiatric resident who had evaluated Mr. Romero and told the resident to "try not to admit the patient" because of the potential disposition problem. At this point the chief psychiatric resident, who had been informed about the case, gave instructions to his junior colleague not to admit Mr. Romero unless he met strict psychiatric admission criteria.

The nursing home was contacted and informed that the patient was stable and could now return. However, the nurse in charge at the nursing home refused to readmit Mr. Romero on the grounds that her staff was unable to manage him. The hospital psychiatrist explained that the patient was stable and did not meet the criteria for admission, and that the hospital could not provide long-term care, but the nursing home still refused.

By then it was 1·00 A.M, and Mr. Romero was sitting quietly in the hospital waiting area, expecting to be transported back to the nursing home, as the psychiatrist had told him. He was not only willing but eager to return to the nursing home. When he was told that he was not being accepted back, he responded that this could not be true and that he would go back by himself. However, he didn't know how to get back. The psychiatrist told him that if he waited until morning, they would probably be able to arrange to have him returned to the nursing home, so he agreed to wait. In order to prevent Mr. Romero from walking out of the emergency room, the psychiatrist placed him on constant observation (known as *one-to-one*) and then left for home. When he returned at 8:30 A.M., he learned that the patient had walked out.

Weeks later, Mr. Romero was picked up in a police precinct in a distant borough of the city. He had been walking around for weeks.

This case illustrates one way in which health care administrators wield increasing power. The battle between the administrative staff at the nursing home and the hospital administrator had little to do with the proper care of a patient and much to do with bureaucratic regulations for admission to a psychiatric unit, rules governing residents in a nursing home, and the convenience of the staff. Mr. Romero's doctor, the psychiatric resident, was powerless to advocate for his patient in the face of administrators who had the authority to decide.

Sometimes hospital administrators are doctors themselves, but their role as administrator typically overwhelms the patient-centered values traditionally inculcated in physicians in medical school and postgraduate training. The nursing home administrator was seeking to dump a patient who had become a management problem, and used as a justification the patient's violation of the nursing home's rules. The hospital administrator desperately tried not to readmit the patient in order to avoid a predictable repetition of the disposition problem this patient had already posed. The chief psychiatric resident tried to comply with the hospital administrator's order by invoking a medically respectable justification: the patient did not fit the criteria for admission to the acute-care psychiatric ward. Even if Mr. Romero didn't really belong in the hospital at this point, the nursing home's refusal to take him back left readmission to the psychiatric ward as the "least worst solution." In the end, there was no one with the power or authority to advocate for the unfortunate Mr. Romero.

What could anyone have done about this situation? As a procedural matter, administrators should not be the final arbiters. When decisions affecting the care of sick, impaired, or vulnerable people have to be made, the needs and interests of the patient should take precedence over other institutional concerns. Even in the absence of an obvious solution to the plight of Mr. Romero, it is clear what the wrong solution was: refusing him readmission to the nursing home and allowing him to wander around the city for weeks.

It is often unclear just how and why hospital administrators enter the clinical picture. In a recent case, a patient with AIDS had been carefully followed at one hospital and was known to the doctors there. The patient arrived suddenly in the emergency department of another nearby hospital in a condition that made the need for resuscitation and intubation an imminent possibility. The patient had previously executed a living will and a do not resuscitate (DNR) directive, which was on file at the first hospital. Both hospitals are part of the same larger academic medical center. When the emergency room physician in the second hospital contacted a colleague from the first hospital, the patient's records were obtained and it was ascertained that a proper DNR order was on file.

At this point, the current physician was comfortable issuing a DNR order to be observed in the emergency room of this hospital. However, a hospital administrator challenged the adequacy of the documentation of the DNR order and stated that the order would not be valid until a copy of the first hospital's papers was faxed to the second facility. (How the administrator got wind of the situation remained a mystery.) The emergency room physician was understandably dismayed at this impediment to appropriate care of patients by physicians seeking to act responsibly. The doctor was denied discretionary authority to accept evidence of a patient's advance directive, transmitted over the telephone by a known and trusted colleague, which was needed in the event of an emergency. If the patient had a cardiac arrest before the fax arrived, the resuscitation team would leap into action, employing all aggressive measures, and the attempt to resuscitate the patient would result in a violation of the patient's rights by ignoring his clearly expressed wishes.

Distant Bureaucrats Apply the Rules

Other forces, operating at a greater distance from the bedside, also operate as constraints on physicians' ability to carry out their obligations to patients. Peer review organizations (PROs) at the state and local levels, state-appointed ombudsmen, and policymakers who devise health regulations regularly intrude into the once sacred physician–patient relationship. They sometimes limit the ability of doctors to act in the best interest of their patients and, more frequently, cause doctors to believe that their power and authority are limited.

The PRO, first developed to protect patients from incompetent or negligent physicians, has become so bureaucratized and rigid that it has the power to second-guess judgments made by physicians on behalf of their patients. This power can result in the imposition of sanctions on physicians who are acting in the best interest of their patients.

Dr. Anton had been Mrs. Goldstein's personal physician for many years. Now in her late eighties, Mrs. Goldstein was severely demented and had a number of medical problems. Laboratory tests indicated that she had iron deficiency, which could be a result of hidden blood loss resulting from cancer of the bowel or of another organ. Dr. Anton wondered whether it was in his patient's best interest to do a workup for iron deficiency anemia. The workup would involve at least one endoscopy, a bone marrow extraction, possibly a barium enema, and possibly an angiogram. All are highly invasive, uncomfortable procedures. Furthermore, given the patient's overall medical status, the information gained from the workup would not really affect the treatment options. The patient could still be given iron for her deficiency.

The physician expressed his concerns about the workup to Mrs. Goldstein's family, and they agreed that the potential risks from and discomfort of these diagnostic procedures would outweigh the benefits to her.

Exercising his best clinical and ethical judgment, and with the concurrence of his patient's family, Dr. Anton decided not to do the workup for iron deficiency anemia. This resulted in a major citation by the PRO. Following a routine review of charts, the PRO assigned "points" (like points for violations on a driver's license) to Dr. Anton, based on their "quality of care" assessment. Although the possibility of appeal does exist and these decisions are sometimes (if rarely) reversed, doctors feel beleaguered by the time and paperwork involved. Their perception is that someone is constantly looking over their shoulder. Physicians like Dr. Anton have begun to feel that their ability to act in their patient's best interest is thwarted by bureaucracy. The end result is that doctors practice defensive medicine, increasing the number of tests they perform, adding to the discomfort of their patients, and, parenthetically, contributing to the escalating costs of medical care.

Federal Government Intrusion: The Gag Rule

Probably the most egregious example of a powerful and distant bureaucracy intruding into the heart of the physician–patient relationship is the federal government's "gag rule," whereby the U.S. government became one of the foremost enemies of patients. This regulation prohibits personnel in health care facilities that receive Title X funds from discussing with a pregnant woman her right to an abortion. It bans all "nonpejorative" speech about abortion in this context.

Congress enacted Title X of the Public Health Service Act in 1970. The act provides federal funding for family planning services but specifically prohibits the use of these funds in programs where abortion is a method of family planning. The intention of that restriction was to limit Title X funds to preventive family planning services, infertility services, and other medical and educational activities.

In 1988, the Secretary of Health and Human Services, appointed by a strongly antiabortion administration, issued new regulations that were designed to provide "clear and operational guidance" to distinguish between Title X programs and abortion as a method of family planning. The new regulations contained three main conditions: (1) a "Title X project may not provide counseling concerning the use of abortion as a method of family planning or provide referral for abortion as a method of family planning"; (2) Title X projects are broadly prohibited from engaging in activities that "encourage, promote, or advocate abortion as a method of family planning"; and (3) Title X projects must be organized so that they are "physically and financially separate" from abortion activities. [10]

The regulation states that Title X projects must direct every pregnant client to "appropriate prenatal and/or social services by furnishing a list of available providers that promote the welfare of the mother and the unborn child." Even if a woman requests referral to an abortion provider, she may not be steered in that direction (except in a medical emergency). The words of the script that the federal government requires physicians in Title X projects to recite are: "This project does not consider abortion an appropriate method of family planning and therefore does not counsel or refer for abortion."

The gag rule thus not only interferes with women's right to receive information about how to procure a safe abortion, it also clashes head on with a patient's moral and legal right to informed consent to medical care. By interfering with the physician's obligation to inform patients about their available treatment options, it prohibits physicians from talking fully and frankly with pregnant patients, from counseling them, and from making appropriate referrals. This federal regulation tears asunder the most basic premise of the physician–patient relationship: that there is complete openness in communication in order to serve the patient's interests.

The 1988 regulations were challenged in a lawsuit filed by recipients of Title X grants and physicians. The suit was filed on behalf of both the clinics and doctors and their patients. They challenged the regulations on the grounds that they were not authorized by the original Title X legislation and also on constitutional grounds—specifically, that the rules violate the Fifth Amendment rights to medical self-determination and to make informed medical decisions free of government-imposed harm, and also the First Amendment rights of patients and health care providers, namely, the freedom of speech clause. After these challenges were rejected by a federal district court and the Court of Appeals for the Second Circuit, the U.S. Supreme Court heard the case known as *Rust v. Sullivan.*

On May 23, 1991, by a vote of 5 to 4, the Supreme Court upheld the Department of Health and Human Services' regulations prohibiting federally funded family planning clinics from engaging in any counseling or referrals related to abortion. The Court argued that because the regulations do not censor *all* doctors in *all* settings, they do not interfere with a woman's right to medical self-determination and to make informed medical decisions. Chief Justice William Rehnquist addressed the concern that poor women do not have access to all doctors or all settings by stating: "The financial constraints that restrict an indigent woman's ability to enjoy the full range of constitutionally protected freedom of choice are the product not of governmental restrictions on access to abortion, but rather of her indigency." The Court's reasons simply ignore the fact that the regulations prohibit physicians or counselors in Title X programs from referring patients to other sources of information where women might obtain uncensored medical advice.

That very circumstance constitutes an injustice, compounding the moral wrongs of intruding into the physician–patient relationship and restricting the rights of the women who use Title X clinics. The injustice lies in the fact that the users of these clinics are primarily poor women, those most likely to lack a personal physician. To deny information and referrals regarding abortion to women who are poor and less well educated is to deprive them unjustly of information readily available to their better-educated and financially better-off counterparts.

Outraged responses to the Supreme Court's ruling in this case came from many quarters. Professor Walter Dellinger, who was co-counsel on a supporting brief for the National Association of Women Attorneys, described the situation as the power of government "officials to order some doctors to provide incomplete or misleading information to pregnant patients."[11] Dellinger observed further that the regulations "require doctors to engage in what would otherwise be actionable malpractice: There are at least 30 states in which physicians can be held liable for failure to disclose treatment options."[12]

An editorial writer assailed the Supreme Court's ruling, saying that it "rests on a perception of reality so skewed as to invite questions about whether five justices have even a common-sense awareness of the world around them."[13] Any woman might construe the doctor's recital of the required statement that the clinic "does not consider abortion to be an appropriate method of family planning" to mean that she cannot have an abortion. Yet, as the editorial notes, "Chief Justice William Rehnquist denies that such enforced advice interferes with the doctor–patient relationship."[14]

In his dissenting opinion, Justice Harry Blackmun pointed out that the regulations "do directly and actively interfere with a woman's right to make the most personal and intimate of decisions, and in some cases, because of the delay engendered by the 'run-around,' might prevent women from obtaining an abortion at all."[15] That result is precisely what was hoped for by the ideologically motivated government officials who promulgated the gag rule. In so doing, they added the U.S. government to the ranks of enemies of patients.

A subsequent attempt by the U.S. Congress to override the ban on abortion counseling in federally funded family planning clinics failed. In November 1991, the House of Representatives and the Senate both voted to knock down the federal regulations. On November 19, President George Bush vetoed the measure. However, there were not enough votes in the House to override the presidential veto, so the regulations remained in force.

The Patient as the Subject of Physicians' Obligation

The traditional ethical portrait of doctors striving to act in their patients' best interest may be oversimplified, if not simplistic. It is easy to cite other roles physicians have had to play, for example, in the context of public health. An ethical requirement to protect the health or safety of individuals other than the patient has dictated a secondary role of physicians: to report communicable diseases, suspected child abuse, and gunshot and stab wounds and to take appropriate steps when a patient is judged dangerous to others as a result of homicidal tendencies or intentional behavior that risks spreading an infectious disease. Like other moral obligations, physicians' duty to their patients, while paramount, has never been absolute.

Now, however, it is not so much the health and safety of the public that are cited as grounds for limiting physicians' overriding obligation to their patients, but rather a vague "duty to society" or a more specific duty to promote values other than health—to minimize the risk of liability, to assist the hospital's cost-control efforts, or to stop wasting resources on "hopeless" patients, those for whom continued treatment is deemed medically futile. These developments threaten to make adversaries out of physicians and patients. Acting in their patients' best interest is only one of several obligations physicians are being asked to fulfill. As members of society, they are being exhorted to reduce costs, to keep an eye on the bottom line, and to place other values above the health and well-being of the individual patient. All together, these developments have contributed to changes in the way medicine has traditionally been practiced.

The growing campaign to ration medical care is endorsed by politicians and legislators and supported by private and government-sponsored insurance providers. Stress on the role of the doctor as society's fiscal gatekeeper does not appear to be integrated into the medical school curriculum in an official manner, but unofficially medical students and trainees are admonished "not to waste society's resources." The message is a mixed one, however, since at the same time, medical students and trainees are taught to employ the most thorough and refined diagnostic procedures, to keep at it until they reach a definitive diagnosis, and to fulfill their obligation as healers to "treat that which is treatable." Physicians at my own medical center who reflect on these issues and take seriously their responsibilities as role models report that they instruct trainees on the prudent use of medical testing as good medical practice, financial considerations aside.

In a different vein, many obstetricians perceive an obligation that threatens the rights of the pregnant women who are their patients. These physicians have a perceived obligation to the fetus as a second patient, with rights and interests that potentially conflict with the decision-making autonomy of the pregnant woman. This rather recent development stems, in part, from a political and ideological battle that has long engulfed this country and has now grown into an unprecedented concern for the fetus, a concern that threatens to overpower respect for the rights of pregnant women.

The prospect that a pregnant woman's fetus may be transformed into her enemy is revealed by scenarios that regularly occur in hospitals. A physician might insist that the woman remain in the hospital for further assessment and monitoring, whereas she feels a greater obligation to her young children in need of her care and attention at home. The hospital administration may seek a court order to override a pregnant woman's refusal of medical treatment, a refusal that would be respected were it made by any other competent adult patient. When physicians seek to override the autonomy of pregnant patients, they may be acting out of genuine concern for their other patient, the fetus, or their motive may be to protect themselves from potential legal liability that could arise if the baby is born damaged. When the hospital administration intervenes to direct the care of pregnant patients, the reason can only be to seek to minimize the institution's liability.

Of course, it is an overstatement to refer to hospital administrators, hospital attorneys, insurance companies, judges, state and local regulators, and sometimes physicians themselves as enemies in any but the loosest sense of the word. I use the term for emphasis rather than precision. Finer distinctions are needed to determine which individuals should properly be viewed as enemies and which are best construed as adversaries or competitors. Sometimes patients and doctors compete for control. At other times, doctors may favor some patients to the detriment of others because of truly limited resources. One patient may then compete with others for those scarce medical resources.

Whatever might be the best term to describe these adversarial or competitive stances between physicians and patients or among patients themselves, they are highlighted by the recognition of patients' rights over the past two decades. What patients have gained in the movement to give them a voice in treatment decisions formerly made by their physicians they have now begun to lose. Individuals occupying administrative roles are succeeding in limiting the rights of patients to determine the nature and extent of medical care. And the professional autonomy long enjoyed by physicians is also shrinking, not only as a result of the enhanced autonomy of their patients but also because of the actions of bureaucratic overseers.

As a supporter of the rights of patients and a defender of patients' autonomy, I tend to reject the paternalistic features of the traditional physician–patient relationship. Those features enabled doctors to conceal things from patients and to take charge of their care, resting on the presumption that the doctor knows best and acts in the patient's best medical interests. Isn't it inconsistent, then, to criticize bureaucratic intrusions on the grounds that they prevent doctors from promoting the best interests of their patients? If the factors characterized in this book as enemies of patients are threatening because they undermine the traditional physician–patient relationship, can one consistently champion the patients' rights movement for changing the way medicine was practiced for centuries?

I believe so. There is no inconsistency in arguing for the importance of recognizing patients' rights and at the same time maintaining that physicians' obligations must be directed toward their patients. Most of the time, patients' rights coincide with their interests, so physicians can simultaneously respect patients' autonomy and act in their best medical interests. On the occasions when rights and interests collide, physicians cannot fulfill both duties at once; but they are still carrying out one of their patient-centered obligations by respecting the patient's autonomy. It is when physicians are compelled by outside forces to act *neither* in their patients' interest *nor* out of respect for their autonomy that they truly become enemies of their patients.

It would be useful to have an account of what the ideal physician–patient relationship should be. Providing a clear articulation of that relationship is not one of the tasks I set out to accomplish in this book, although I can imagine critics saying that without that clear account, the rest of what I say here is muddled. I can defend my limited aims by reiterating first that physicians' twin obligations—to promote their patients' best interests and to respect their autonomy—are central and normally do not conflict with one another. Second, the "enemies" depicted in ensuing chapters are frequently the enemies of physicians as well as of patients. Hospital administrators, hospital attorneys, risk managers, and bureaucrats who sit somewhere reviewing charts, limit physicians' professional autonomy and often compel doctors to act in ways that are contrary to their patients' interests as well as violative of their rights.

Finally, the fact that we can conjure up a notion of the ideal physician–patient relationship does not mean that the ideal was ever realized historically. But to argue that there never was a time when doctors strove unceasingly to serve their patients and patients responded with gratitude is not to deny the value of the ideal. The distinction between what is (or was) and what ought to be must be preserved if people are to reason and act from an ethical point of view.

Notes

1. Quoted in President's Commission for the Study of Ethical Problems in Medicine and Biomedical and Behavioral Research, *Making Health Care Decisions* (Washington, D.C.: U.S. Government Printing Office, 1982), vol. 1, p. 32.

2. Ibid., p. 96.

3. Approved by the American Hospital Association's (AHA) House of Delegates, February 6, 1973. The statement was published in several forms, one of which was the S74 leaflet in the Association's S series. The statements quoted here were copyrighted in 1975 by the AHA.

4. The writer was a reviewer of a prospectus I submitted for this book, whose identity is unknown to me.

5. Ruth Macklin, *Mortal Choices: Bioethics in Today's World* (New York: Pantheon, 1987), pp. 163–64.

6. Marshall B. Kapp, "Are Risk Management and Health Care Ethics Compatible?" *Perspectives in Healthcare Risk Management*, vol. 11 (Winter 1991), p. 3.

7. Ibid., p. 4.

8. Ibid.

9. This case was presented by a psychiatrist at a monthly conference on ethics, law, and

psychiatry. The patient's name has been changed to protect confidentiality in the description of this case and all others throughout the book.

10. *The United States Law Week*, Extra Edition No. 1, Supreme Court Opinions, Vol. 59 (Washington, D.C.: Bureau of National Affairs, May 21, 1991), p. 4453.

11. Walter Dellinger, "Conservatives Play Doctor," *New York Times* (May 25, 1991), Op-Ed page.

12. Ibid.

13. John P. MacKenzie, "What the Doctor Ordered," *New York Times*, Editorial Notebook (June 3, 1991), Editorial page.

14. Ibid.

15. Ibid., pp. 7–8.

2

Law as an Advocate for Patients: A Case Study of DNR

One way of trying to ensure that the moral rights of people are adequately protected is to back them up with laws. Patients' rights have been recognized through court decisions, such as the series of rulings that mandated a patient's informed consent to treatment and the right to refuse life-sustaining therapy; through legislation, as in the living will laws that have now been enacted in all of the states; and by state or federal regulations, such as those governing research involving human subjects. It is ironic but true that sincere efforts to change a bad situation can lead to unintended consequences, ones that raise questions about whether that situation has actually improved. This is what happened in the attempt to change an ethically problematic practice surrounding orders not to resuscitate some patients who suffered a heart attack.

Cardiopulmonary resuscitation (CPR) poses some peculiar problems. A resuscitation team must act swiftly and surely, pounding on the patient's chest, inserting a breathing tube, administering injections, and monitoring the patient's vital signs. On the one hand, CPR is a physically aggressive procedure that can cause serious harm, such as breaking many bones in the chest of an elderly patient. On the other hand, since it is a life-sustaining— or, more accurately, a life-restoring—procedure, it promises the benefit of continued life. It is generally assumed that most reasonable people would wish to have their hearts restarted or their breathing restored if they experienced a cardiopulmonary arrest. However, it turns out that some patients who are resuscitated are never again able to breathe on their own. Quite a few reasonable people have judged that a life that can be prolonged only by their being permanently hooked up to a respirator is a life they do not wish to endure. The issue is further complicated by the fact that CPR is an emergency procedure, and therefore that patients in need of resuscitation cannot be consulted about its risks and benefits at the time it must be performed.

This medical procedure thus raises a rather unique set of questions: Is CPR a treatment for which patients should be required to give their informed consent? Or should it be assumed that patients would consent, if they were able, so that consent must be obtained for *not* administering CPR? Is it even necessary to obtain the patient's consent for either administering CPR or withholding it, given the fact that it is an emergency procedure and doctors need not obtain informed consent in an emergency?

Furthermore, as is true of any medical procedure, patients and physicians may disagree prospectively about whether CPR should be performed. A physician might judge CPR to be contraindicated and suggest that a DNR order be written, but the patient or the family may disagree. Conversely, the patient or family might wish to forego CPR based on the patient's quality of life, but the physician may disagree based on a felt obligation to prolong life when a patient is not terminally ill. Finally, from fear of liability, the hospital administration might insist that all patients be resuscitated, even in cases where patients, the family, and the physician all agree that CPR should not be performed. This problematic situation gave rise to pleas for an ethically satisfactory response.

The "Do Not Resuscitate" Order

When it was first developed, CPR was considered an emergency intervention for people who suffered a cardiac arrest or suddenly became unable to breathe on their own. As this life-restoring procedure was refined, it became part of the normal routine of acute care hospitals.[1] Once a medical procedure becomes routine, medical personnel are normally prepared to administer it, and patients and families come to expect it. That this treatment could be beneficial was undeniable. Especially in circumstances where the cardiac arrest was unexpected, or occurred during or following surgery, patients could be brought back to life and often restored to their previous level of functioning.

Hospitals developed various "codes" to designate the situation in which emergency teams were mobilized to perform CPR. One common term used was "Code Blue." One hospital in New York City used a three-digit code that was originally an emergency telephone number, 1-2-6, to designate the need for emergency resuscitation, and another used "Code 33." Other hospitals used the loudspeaker paging system to announce: "Calling Doctor Pacemaker! Doctor Pacemaker to 8 South!" Physicians and nurses used the expression "call a code" to refer to the initiation of CPR, which became shortened to "the patient was coded," meaning "was resuscitated." This eventually shifted to a general practice of using the term "coded" as a verb in place of "experienced a cardiac arrest" (as in "the patient coded").

Even more creative than the terminology were the methods developed to designate which patients were not to be resuscitated. As experience grew in resuscitating patients with all sorts of diseases, including many with terminal illnesses or conditions in which CPR could be truly burdensome, it became evident that full resuscitative efforts were sometimes an inappropriate response to a cardiac arrest. Not every patient could benefit from being resuscitated. Some could only be revived temporarily, and their hearts would soon stop again. Others would suffer damage from the resuscitative procedure itself, such as broken bones from pounding on the chest. Still others might never be able to breathe on their own after CPR was initiated and would have to remain permanently on a respirator. One author notes that "Many of the procedures are obviously highly intrusive, and some are violent in nature. The defibrillator, for example, causes violent (and painful) muscle contractions which . . . may cause fracture of vertebrae or other bones."[2]

In such circumstances, it became evident that resuscitation need not and should not be initiated every time it was possible to do so. A new form of doctors' orders was devised: the "Do Not Resuscitate (DNR)" order. Beginning in 1976, articles began to appear in the medical literature describing the policies and practices on "no code" orders that were developed in some hospitals.[3]

A study carried out in 1981 in Boston's Beth Israel Hospital posed, among others, the following questions: How often and under what circumstances do physicians discuss CPR with patients or their families? To what extent do physicians form attitudes about resuscitation without direct communication with the particular patient? When are families, but not patients, consulted? How does the physician's perception compare with the patient's stated preference about resuscitation? Is there a discrepancy between what physicians say they ought to do and what they actually do?[4]

Not surprisingly, the research revealed that only a small percentage of patients had been consulted about resuscitation before a cardiac arrest by either their private physician or a house officer or both, while families were consulted in a much greater number of cases. Even those physicians who believed that patients should participate in decisions about resuscitation admitted that rarely did they initiate a discussion of the issue with patients. Of 24 patients with decisional capacity who survived CPR, the preference they expressed correlated only weakly with their physician's opinion about their desire for resuscitation.[5]

Writing a DNR order made many physicians uncomfortable. Some doctors were adamantly opposed to the very existence of DNR orders in the belief that the physician's obligation is always to try to save a patient's life, especially in an emergency. At one conference in which attending physicians and trainees were gathered to discuss a newly established DNR policy in the hospital, a surgeon waved the policy angrily, saying, "This is a death sentence!" However, most physicians recognized at least some circumstances in which resuscitation would not truly benefit the patient. But they rarely discussed the DNR order with their patients.

Physicians typically (and perhaps understandably) resist holding discussions with patients that bring up unpleasant future scenarios. Once CPR was established as a standard procedure for patients who experience cardiac arrest, doctors became reluctant to discuss with patients whether or not they wished to be resuscitated. That silence might be defensible if resuscitation were always in a patient's best interest and so long as all patients were being resuscitated. But in many cases, doctors judge resuscitation not to be in the patient's best interest. For example, the patient may be dying of cancer with widespread metastases, and cardiac arrest may signify her body's inability to sustain life any longer. To bring the patient back to life with CPR would only prolong a painful dying process. Alternatively, the patient may be very old and frail—say, in her nineties—and aggressive CPR would result in breaking all her ribs, even if it succeeded in resuscitating her heart for a short time. In less obvious situations, the doctor may simply view the patient's quality of life as so poor that continued life would not be of much benefit.

A widespread practice grew up around the writing of orders not to resuscitate. After determining that resuscitation was not in the patient's interest, and without consulting the patient, the doctor would write a DNR order in the patient's chart. Physicians who wrote such orders without holding a discussion with their patients sought to justify their actions by referring to the "therapeutic privilege" to act in the patient's best interest. Only this time, what is said to be in the patient's best interest is to be spared the burden of discussing these troubling matters. As a result, a practice grew up of writing DNR orders without regard to patients' right to participate in decisions about their own care and treatment. Although this practice has been changing in the past few years, it illustrates the view still held by some doctors that patients' rights often stand in the way of their best interests. If holding a discussion about resuscitation with a patient is likely to cause the patient emotional discomfort, these doctors maintain that their primary obligation is to "do no harm," an obligation only likely to be thwarted by honoring the patient's rights.

The circumstances that led to passage of a New York State statute pertaining to DNR orders illustrate how patients' rights can conflict with their own best interests, and also how doctors have struggled to maintain decision-making authority in a changing era of medical practice.

The remedy adopted in New York was enactment of detailed legislation, which specified precisely the rights of patients and the corresponding obligations of physicians and hospitals regarding DNR orders. Although many would argue that passage of this law has been a victory for patients' rights, others contend that the law imposes too many constraints on physicians and, furthermore, has failed in its aim of safeguarding patients' rights and interests. These detractors argue that legal and regulatory interference in the physician–patient relationship only has the effect of undermining the relationship by bureaucratizing key elements of medical practice. Judging the validity of this criticism requires a closer look at what led up to passage of the law and what has happened since it was enacted.

Chaos and Dishonesty

Hospital administrators and risk managers were in a perpetual quandary about what stance to take regarding orders not to resuscitate. Some hospitals urged their staff never to talk about DNR orders to anyone outside the hospital, especially journalists or reporters. I was told by one hospital's risk manager that if any reporter ever asked me whether the hospital had a DNR policy, I shouldn't reply. Other administrators insisted that no policy be put in writing. To do so would constitute written proof that some patients were being allowed to die without any effort being made to resuscitate them. In contrast, another group of hospital attorneys insisted that the risk of liability was greater if there was no written policy than if a clear statement existed, outlining the hospital's procedures regarding DNR orders. Since everyone was aware that DNR orders were used in hospitals, it was better, they argued, to put the matter up front by having a well-drafted policy consisting of sound procedures.

The uncertainty and chaos surrounding DNR orders were captured in testimony by one physician from a community hospital at a meeting of the President's Commission:

> Older physicians are afraid of putting "do not resuscitate" down because they are
> afraid of being sued for making a wrong decision. The younger physicians are
> anxious to put a "do not resuscitate" down because they are afraid of making a wrong
> decision. The nurses will not act without a "do not resuscitate" because they are
> afraid of being sued.[6]

Why did such a fuss arise concerning this particular medical order, and why were there so many conflicting opinions? Two features, taken together, add up to an idiosyncratic situation. The first is the fact that unlike other medical orders, an order not to resuscitate is an order *not* to carry out a medical procedure. Although the procedures themselves are invasive and aggressive, patients entering a hospital were assumed to consent to resuscitation unless there was a specific order to the contrary. The second feature is the fact that these orders were frequently written without consultation with the patient and, somewhat less often, without even the knowledge or consent of the patient's family. It remains a mystery just what patients knew about the existence of DNR orders in general, about the components of cardiopulmonary resuscitation, or about what was to take place if they suffered an arrest during their hospitalization. Their ignorance was a consequence of physicians' failure to discuss resuscitation or DNR orders with them.

In many cases, patients lacked the capacity to participate in a DNR discussion. It was easier for doctors to have that discussion with family members than with those patients who still maintained decisional capacity, and in some instances such discussions took place. Nevertheless, a lawyer with a New York City law firm wrote in a letter (dated Jan. 10, 1983) that "Few DNR orders are being written for incompetent patients in New York State at present, largely because district attorneys state that they consider such orders to be illegal and subject to criminal prosecution."[7]

But there were also many cases in which physicians failed to hold a DNR discussion with the patient's family and the order was written anyway. In those circumstances, a deceptive practice known as a "slow code" or "show code" began to emerge. If a patient had been made a DNR without either the patient's or family's consent and the family was present when a cardiopulmonary arrest occurred, the staff felt that they had to act as if they were doing something to revive the patient. Their fear, somewhat vague and often unarticulated, was that the family would be horrified at the inaction, and that they might sue the doctors and the hospital. Whether or not those fears were justified, it was certainly true that writing DNR orders without the patient's or family's participation violated a deeply rooted ethical principle and probably also existing laws mandating informed consent.

This serves as a reminder of the peculiarity of DNR orders. In other situations, the patient's consent is required before an invasive procedure may be undertaken. As long ago as 1914, U.S. Supreme Court Justice Benjamin Cardozo ruled that for a surgeon to perform an operation without the patient's consent constituted an assault. But surely, *not* performing a medical procedure could not constitute an assault. Why should the patient's consent be required not to do something aggressive and invasive? The answer is that cardiopulmonary resuscitation had become a routine procedure in hospitals. In acute care settings, where resuscitation equipment is available and medical and nursing personnel are trained to do the procedure, it is now a component of standard care. Therefore, to omit or forgo a procedure considered routine would be to deviate from the established standard of care. The principle of respect for autonomy mandates that patients must be consulted when physicians propose a nonstandard treatment plan.

But physicians recoiled from discussing DNR orders with their patients. The very discussion could harm the patient, they argued: As physicians, they were obligated to do no harm. It would surely make some patients depressed to realize that resuscitation would likely carry more burdens than benefits. That would be harmful. Also, a patient's medical condition might be worsened if the patient were forced to discuss a worrisome, unpleasant topic. Besides, some physicians contended, it was also their duty to help the patient maintain hope. These claims by physicians are long familiar in the broader context of informed consent and disclosure to patients about their illness. But in the 1980s, when the issue of DNR orders became prominent, it was too late for these paternalistic defenses to carry much weight.

Ethicists working in hospitals deplored the fact that physicians were unwilling to have DNR discussions with their patients. Biomedical ethics has from the beginning championed the cause of patient autonomy, and ethicists asked: Don't patients have the right to participate in decisions about resuscitation, just as they do about any other decision about medical treatment? Hospital ethics committees criticized the lawyers and risk managers who were reluctant to establish policies requiring the patient's consent before a DNR order could be written. Fearing legal liability, few hospitals would acknowledge that they had policies, even when they did exist. And perhaps their fears were not entirely unwarranted. At one professional society meeting held in New York City in 1982, a county prosecutor proclaimed that if he learned of any doctors or hospitals that countenanced DNR orders, he would move to prosecute them for murder. He said that as he discovered them, he was drawing up a list of facilities in which DNR orders were being written. This chilling episode shows that the fears expressed by risk managers are not always paranoid. But their response to those fears—a conspiracy of silence and denial—was ethically unconscionable.

Meanwhile, among hospital personnel, furtive practices were the rule when it was decided to label a patient DNR. In addition to the slow codes (the resuscitation team walked through the procedures) and the show codes (sometimes termed "Hollywood" codes), secret codes were used to designate patients who were no codes. In one hospital, the names of patients who were given no-code status were marked each night with an asterisk on a blackboard. A hospital in Queens, a borough of New York City, used a system of affixing stick-on purple dots to file cards, kept in the nursing record, of patients who were made DNR. The cards were not routinely seen by physicians in the hospital, and were discarded when the patient died or was discharged.

By the end of 1982, the Medical Society of the State of New York decided to issue guidelines for withholding emergency resuscitation. Yet even with these guidelines, issued by a prestigious medical organization and distributed to the state's two major hospital associations, fear of liability persisted in hospitals. Most hospital spokesmen continued to deny either that DNR orders were being written or that the hospital had a written or unwritten policy. The Medical Society's step was viewed by many as mere guidelines, which did not require adherence.

Purple Dots in Queens

At about the same time, the secret purple dot system at LaGuardia Hospital in Queens was under investigation by the New York Attorney General's office. A lawyer for the hospital denied that the institution had done anything improper.[8] When a special grand jury found that "shocking procedural abuses" had occurred, the same lawyer called the charges in the grand jury report "the height of irresponsibility." He denied that the hospital had any DNR policies or purple dots.[9]

The Grand Jury Report in the LaGuardia Hospital case did not fault physicians for writing DNR orders. On the contrary, the report acknowledged circumstances in which cardiopulmonary resuscitation would be inappropriate. The concerns expressed in the report centered on the actual and potential abuses, given the way DNR was being handled. The report urged the state "to insure that the decision-making process is governed by explicit procedural safeguards to prevent the decision from being made carelessly, unilaterally or anonymously."[10] The grand jury emphasized the need for securing the patient's "knowing and informed consent," and for making sure that a qualified physician was involved in evaluating the patient's condition.

Members of the hospital administration testified at the grand jury hearings in the LaGuardia Hospital case. They insisted that the prohibition they had issued against writing such orders "was justified by uncertainty over the legality of 'no-coding.' "[11] The grand jury found that hospital officials were the ones who instituted the purple dot system in order "to avoid leaving any tangible evidence of 'no-code' orders."[12] The executive vice-president of the hospital was an attorney, whose testimony before the grand jury confirmed that his request not to put a formal note in patients' charts was intended to minimize legal exposure. The report stated that "No medical or ethical justification for not putting 'no-code' orders into written form was ever suggested."[13] In the end it was a nurse, the associate director of nursing, who distributed the " 'purple dots'—small, colored, commercially available, adhesive-backed paper discs—which she and other supervisors instructed the nurses to attach to the 'Kardex' file of patients whom the doctors orally designated as not to be 'coded.' "

The Grand Jury Report in the LaGuardia case reveals that it was hospital administrators, physicians, including the chief of surgery, nursing administrators, and others in positions of responsibility and authority who devised and implemented the infamous purple dot system. But the report also notes that many nurses in the hospital were deeply concerned over the absence of any DNR standards in the institution. For example, they recounted episodes in which a doctor on the day shift wanted a patient coded, while a doctor on the night shift treated the same patient as a no-code. In addition, because of the haphazard way these orders were issued, there was no review procedure and no formal method for revoking a DNR order if the patient's condition improved. Some of the nurses criticized the system as being insecure, since many people had access to the Kardex files. Nurses also expressed the worry that the dots could be put on the wrong card, or fall off and stick on another card to which they did not belong. [14]

That particular worry was apparently well founded. A nurse kept one card, a file of a Medicaid patient who had died at the hospital, that had two purple dots affixed to it. The grand jury could not ascertain how either dot had gotten onto the Kardex file, since all the physicians who played a major role in treating that patient denied knowledge of the dots. Nurses at the hospital were troubled by the fact that "they were responsible for documenting the decisions which the doctors were unwilling to put in writing."[15] The nurses, then, were the ones who would have been held accountable, since they were the ones who actually put the dots on the patients' file cards. [16]

The Grand Jury Report in what came to be known as the "purple dot case" was widely publicized by the media. Similar practices were going on in other hospitals at the same time. On March 28, 1984, the *New York Post* carried the headline "Prosecutor Demands Hospital Action: 'Ban Secret Death Lists.'" The newspaper article referred to "blackboards of doom," a system allegedly used at the prestigious Sloan-Kettering Hospital to mark the names of patients who were or were not to be resuscitated with code letters. The *Post* reported that hospital officials "vehemently" denied that the procedure was used to destroy evidence that might be used in malpractice suits against doctors or the hospital. Yet one physician from Sloan-Kettering did admit that physicians did not discuss the code classifications with the patients whom the letters designated.

All of these events led to a call for legal recognition of DNR orders in the State of New York. The President's Commission stated the need for explicit hospital policies on the practice of writing and implementing DNR orders but stopped short of proposing the need for legislation or government regulation. [17] However, the Grand Jury Report in the LaGuardia Hospital case said that the state "can and should establish such procedural safeguards."[18] The first recommendation in the Grand Jury Report reads:

WE RECOMMEND that the State Legislature enact basic standards to regulate the procedures by which physicians decide whether cardiopulmonary resuscitation should be withheld from terminally ill patients, and authorize and direct the Commissioner of Health to promulgate specific regulations to implement and enforce such standards. [19]

As informal efforts to find a solution to the complications surrounding DNR orders had apparently failed, a formal legal solution was now proposed.

The New York State DNR Law

In April 1986, the New York State Task Force on Life and the Law, a commission appointed by the governor, issued a report and proposed legislation on DNR orders. Based on that report, the New York State legislature enacted a statute that went into effect on April 1, 1988. Although most states had enacted some form of legislation regarding living wills or appointing a health care agent or proxy, New York was unique in having enacted a law specifically devoted to DNR orders. This was even more surprising in light of the fact that the state had at that time no living will legislation or statutes addressing the more general question of withholding life-sustaining treatment from patients. The reason is largely a historical one, stemming from the purple dot case and the ensuing Grand Jury Report, but there were sound ethical underpinnings for the law.

First and foremost, the law is designated to ensure that patients with decisional capacity be full participants in a DNR discussion and subsequent decision. For patients who lack decisional capacity and have not previously indicated their wishes about resuscitation, family members are authorized to consent to DNR orders but only after certain medical criteria are satisfied. The most stringent safeguards apply in the case of patients who lack decisional capacity and who have no family member or significant other to consent to a DNR order. The only condition under which a DNR order may be written for such patients is that in which resuscitation is deemed medically futile. The law defines "medically futile" as follows: "Medically futile means that cardiopulmonary resuscitation will be unsuccessful in restoring cardiac and respiratory function or that the patient will experience repeated arrest in a short time period before death occurs.[20]

Based on the experience of other states, which lack specific laws regarding DNR, the need for such a law might be questioned. However, since the law in New York went into effect, a more pressing question has been frequently asked: is it desirable to have such a law? The legislation was enacted to prevent abuses, to protect the right of patients to participate in DNR decisions, and to ensure their well-being. But a curious complication has developed. A set of complex regulations accompanying the law has led some observers to suspect that doctors are refraining from writing DNR orders altogether, thus undermining one purpose the law was meant to serve.

One reason cited for physicians' avoidance is the sheer amount of paperwork involved. One hospital has devised a series of forms to be filled out in connection with DNR decisions. A 35-page packet of materials distributed to chiefs of service, attending physicians, and resident physicians includes a covering memorandum, a flow sheet to help physicians identify which of six different categories applies to a particular patient and which of six different documentation sheets should be used, along with the six documentation forms and a number of subsidiary documents.

DNR Documentation Sheet 1 is for Adult Patient with Decisional Capacity. Sheet 2 is for Adult Patient, Therapeutic Exception, for cases in which the physician believes that the patient would suffer immediate and severe injury from a discussion of CPR and wishes to bypass holding a DNR discussion with the patient. A concurring physician's signature is required on this form, attesting to the first physician's determination "to a reasonable degree of medical certainty that the patient would suffer immediate and severe injury from a discussion of CPR," along with the reason on which the determination is based. Sheet 3 is for Adult Patient without Capacity Who Previously Consented to a DNR Order. Sheet 4 is for Adult Patient without Decisional Capacity and with a Surrogate Who Has Not Previously Consented to a DNR Order. Sheet 5 is for Adult Patient without Decisional Capacity Who Has Not Previously Consented to a DNR Order and without a Surrogate. Sheet 6 is for a Minor Patient (the patient may have a major illness but is under 18 years of age). And then there are additional pages for Consent by a Surrogate to DNR Order, Affidavit of Close Friend (which must be notarized), and subsidiary forms to document a patient's lack of decisional capacity, concurrence by another attending physician, a notice to the patient and the surrogate of lack of decisional capacity, and a form to document that a patient lacks a surrogate. It is little wonder that physicians consider the process set in motion by the DNR law to be burdensome.

A second reason for physicians' tendency to avoid writing DNR orders is their uncertainty or misunderstanding of what the law actually says. Rather than risk violating the law when its provisions are unknown or misunderstood, some physicians do nothing. A frequently asked question is: Does the law require CPR to be performed on *every* patient for whom a DNR order has *not* been written, even if it would clearly be futile or detrimental to the patient? This question was answered by Robert Swidler, who served as staff counsel to the New York State Task Force on Life and the Law.[21]

Swidler notes that some medical personnel have expressed grave concern about what they believe to be a requirement to resuscitate all patients for whom a DNR order has not been written. Some doctors even contend that it compels them to violate their oath to do no harm. By contrast, Swidler maintains that "the view that the presumption of [a patient's] consent [to treatment] constitutes an absolute duty to resuscitate is at odds with what the statute says, what the drafters and legislature intended, and what good medical practice requires."[22]

What, then, is the proper reading of the presumption of consent? The correct interpretation of the law is that the non-DNR patient (a patient for whom a DNR order has not been written) is deemed to *permit* resuscitation. This contrasts with a reading of the law that would *require* physicians to resuscitate all non-DNR patients. The duty to provide resuscitation to a non-DNR patient is based on the applicable standard of care that governs the conduct of health care professionals and facilities. This interpretation of the law no doubt makes more sense to the lawyer who assisted the Task Force in making the DNR recommendations than it does to physicians engaged in everyday practice.

What is the standard of care regarding provision of CPR? The application of the standard of care to any medical treatment is necessarily inexact. But the outer boundaries of the duty are clear. In particular, "if a physician makes an on-the-scene decision to withhold CPR from an arresting patient who does not have a DNR order because of a well-grounded judgment that CPR at that juncture would be medically futile, the physician will not be liable for malpractice or subject to other sanction: he or she did not omit a measure that an ordinary doctor would provide."[23]

There are a number of other misconceptions and frequently debated points about what is required by the DNR law in New York State. One implication of the law should be made clear: physicians who feel burdened by the requirements of the DNR law to assess the competent patient's wishes *may not* simply await the patient's arrest and then direct the withholding of resuscitative efforts on grounds of futility. Such deliberate attempts to dodge the law were specifically prohibited in a Health Facilities Memorandum issued in March 1988 by the New York State Department of Health. The central role of the DNR law is to ensure that DNR orders are properly documented and that they are based on the consent of the patient or an appropriate surrogate decision maker.

Although framers of the law contend that it does not increase, decrease, or alter existing standards about when in the absence of a DNR order CPR must be provided,[24] it is apparent that physicians are still either ignorant or confused or both regarding those standards. They are also ignorant of specific ingredients of the law, such as the definition of "futility" supplied by the statute. House staff (interns and residents) continue to use vague and subjective definitions of futility, including their perceptions that the patient is too old or too demented to benefit from resuscitation or their prediction that the patient "won't live to walk out of this hospital."

A third reason for avoiding DNR orders is physicians' continued reluctance to bring up the subject of death with patients who still have decisional capacity. Requiring doctors to hold discussions with patients about cardiopulmonary resuscitation was one of the chief purposes the law was meant to serve. It was intended to prevent the writing of DNR orders without the knowledge and consent of patients who are able to participate in these decisions. Yet many physicians still feel uncomfortable initiating such discussions, so under the current legal mandate they avoid the discussion and refrain from writing the order.

Like other laws and regulations, the DNR law has some rigid provisions. For example, it specifies a precise order of individuals who may act as surrogates for the incapacitated patient. If the patient, while still competent, had designated a surrogate to make a DNR decision, the law requires that physicians honor the decision of the patient's surrogate. That part of the law is fine, since it mandates respect for the patient's own previously expressed choice. But if the patient has not delegated a surrogate, then the law lists in a specific hierarchy family members who must be consulted. Even if someone further down the list knows the patient better and could therefore make a decision more in line with the patient's own values, a relative higher on the list must be consulted first. The law does have an innovative feature, however, in that it includes on the list not only relatives but a close friend who may be called upon to speak for the patient.

Although this hierarchical list of potential surrogates has the shortcoming of being inflexible, there is no alternative that would in general be superior as a method of choosing a surrogate. This aspect of the DNR law has caused the most difficulty in cases involving AIDS patients. The person who is usually closest to a gay man suffering from AIDS dementia, and most in touch with his current values, is his lover. But if the patient never designated his lover as his surrogate, and his parents come forward, the law recognizes the parents rather than the lover as the only ones who may make a DNR decision on behalf of the patient. The wishes of parents, who might have been estranged from their homosexual son, often conflict with what the son's companion says the patient would want in this situation. It is probably true in most cases that a gay man's current lover is better acquainted with his wishes about resuscitation. Yet for their part, the parents may still make a DNR decision that they sincerely believe is in the patient's best interest.

Physicians' Responses to the DNR Law

Anecdotal accounts and published reports suggest that many physicians view New York's DNR legislation as a step backward. Dr. Catherine Hart, an internist at New York Hospital, said: "the new procedure is expensive, cumbersome, and unnecessary." Dr. Leslie Kohman, a cardiothoracic surgeon in Syracuse, complained that "now we have to fill out a seven-page document and get many signatures, which is a waste of time. It harasses doctors and families." Dr. Thomas Cardillo, an internist in Rochester, contended that "we have taken an issue that belongs in a private discussion and subjected it to law. Hospitals want regulations that place more of the burden on doctors and patients. The law's intention is good, but it's like legislating morality: It just can't be done."

And Dr. Frank Brescia, medical director of Calvary Hospital in the Bronx, asserted: "I think it is terrible. In some ways it puts the patient's right of autonomy and the physician's right of doing what is best for the patient in conflict. In spirit, the law should allow people to die. But DNR will have a reverse effect. It is easier to resuscitate than it is to follow the law and bombard the family with the law's requirements."[25]

One surgeon expressed to me his view that "the DNR law is a bad thing— it's irrational and unnatural." The surgeon said it was irrational because it separates resuscitation from all other medical procedures. It would have been better, he stated, if the legislature had passed a comprehensive law stipulating criteria and procedures for withholding all other critical or life-preserving medical procedures in addition to CPR. "We were better off with purple dots," the surgeon mused.

This doctor also pointed to an unwanted consequence of the law: it has led to actions neither foreseen nor intended by the framers of the law, such as refusing to admit patients with a DNR order to an intensive care unit (ICU) or removing them from the unit to a back room once a DNR order is signed—two responses that have become common practices in some hospitals. ICU personnel argue that consistency of treatment demands that a patient who occupies a bed in a critical care unit should be treated aggressively for every medical indication. A patient with an order not to resuscitate simply does not belong there. However, the statute clearly specifies that a DNR order is consistent with the provision of other life-sustaining treatments. The patient may consent to a DNR order and at the same time express the desire for other life-prolonging measures. Intensivists (specialists in critical care medicine) regard that position as irrational. One remarked, "That's what happens when lawyers and legislators start meddling with the details of medical practice."

An even more disturbing trend is that of relegating DNR patients to a section of a floor or a unit in the hospital where only minimal monitoring and care are performed. The surgeon who told me he thought the DNR law was a bad thing expressed this practice by the formula: DNR = DNRx. In a different hospital, which has a minimal care section known as the "Drei" unit,[26] when a DNR order is written for a patient the house staff appends an informal note with the initials "TDTD," which stand for "To Drei to Die." These new developments reveal a resistance to change on the part of doctors and a tendency for practices to grow up in response to mandates that physicians feel are intruding into their medical prerogatives or are otherwise unacceptable.

The reverse situation also occurs, stemming from a misunderstanding of how the DNR law is properly to be applied. In one case, members of the nursing staff were opposed to a medical recommendation not to give nutritional supplementation to Mrs. R, an elderly woman dying of colon cancer. The nurses said it was "illogical" not to insert a nasogastric tube to begin artificial feeding when no DNR order had been written for the patient. They argued that without a DNR order, there was no basis for not feeding the patient. But nothing in the law suggests that other life-prolonging treatments may not be withheld if a patient is a candidate for resuscitation. In this case, the oncologists in charge of the patient's care judged it to be inappropriate to institute artificial feeding, given Mrs. R's terminal condition and her mental confusion. Although the oncologists' judgment not to institute artificial feeding might be questioned on other grounds relating to the patient's condition and prognosis, it is a mistake to question their decision simply because the patient had not been made a DNR.

Nurses frequently object to withholding artificial nutrition and hydration, claiming that it is wrong to "starve patients to death." The debate surrounding the ethics of forgoing food and fluids is legitimate but should be kept separate from the issue of orders not to resuscitate a patient. In the case of Mrs. R, an additional factor motivated the nursing staff's disapproval of withholding nutritional supplementation. The nursing supervisor claimed that if the patient were sent back to the nursing home where she had been before this hospitalization with more than a two-pound weight loss, the hospital would be cited by the state regulator reviewing the charts. Fear of a citation by regulators was thus an added reason for the nurses' opposition to the medical plan, a reason that had nothing to do with the well-being of the patient.

The experiences of physicians, house officers, medical students, and ethicists in numerous hospitals in New York demonstrate that the existence of a statute on orders not to resuscitate has not eliminated the slow codes and show codes prevalent in the past. This practice occurs regularly in cases where a patient who has lost decisional capacity had never signed a DNR order and the patient's family refuses to sign. If the physicians believe that the patient is not a proper candidate for resuscitation despite the absence of a DNR order, a code is called and they simply go through the motions. Drawing the curtain, they shout out the orders that would be followed in the case of a genuine code. To the family in the corridor, it sounds as though a resuscitation is being performed. When the resident calls for a syringe, he squirts its contents into the air. The patient dies, and the family is told the patient couldn't be saved.

Some of these Hollywood codes are done in direct violation of the law. For example, an attempt at resuscitation would not be futile in the sense defined by the statute but the physicians think resuscitation is contraindicated, the family has been asked but refused to consent to a DNR order, and so the house staff simply goes through the motions to satisfy the family. Other show codes are performed in ignorance of what the law actually requires. For example, genuinely futile attempts are made that are not required by law, but rather because the house staff and the nurses mistakenly believe they are legally mandatory.

Still other show codes are done because of the peculiarities of the differing responsibilities of physicians and nurses. Nurses are responsible for calling the code, but not for the assessment of the patient's medical condition. It is up to a physician to ascertain whether resuscitation would be futile, in strict adherence with the legal definition of the term. So if a patient without a DNR order experiences a cardiac arrest, nurses feel obligated to call a code, whatever the patient's medical condition. Interns and residents, arriving on the scene to begin the resuscitation, judge the patient to be too far gone already, or they know from their experience in caring for the patient that resuscitation would be futile. However, the code has already been called, and they can't just stand there; so, they go through the motions. Whatever the reasons, and whether or not they are in direct violation of laws or hospital policies, show codes are deceptive practices, deliberately designed to fool someone into thinking that patients are being treated aggressively when they are simply (perhaps appropriately) being left to die.

Observations such as these, along with case examples from different hospitals, suggest that this well-intentioned and much-needed law has encountered unanticipated problems, with the likely result that at least some and perhaps many patients are failing to derive the intended benefits. The law was intended to institute safeguards, prevent abuses, and in general ensure that cardiopulmonary resuscitation is appropriately carried out or withheld. The following cases provide a more detailed indication that these intended purposes are still being thwarted.

Case 1: Medical residents convinced the family of an 86-year-old man that the patient should be made a DNR. At first, the family had requested resuscitation, but the residents argued for a DNR order. The attending physician on the ward told the family (the patient's wife, son, and daughter) that he didn't think the patient should be a DNR. The attending physician said he wouldn't be able to continue on the case if the family insisted on a DNR order. However, that may well have been a reflection of the physician's personal values and religious beliefs regarding DNR orders rather than a medical judgment that the patient could truly benefit from resuscitation.

The family had come to trust the house staff, with whom they had more contact than with the attending physician during their visits to the hospital, and they went along with the resident's view of the appropriateness of the DNR order. It emerged that the process by which the order was written was questionable, and the letter of the DNR law was violated.

A medical student who presented this case at ethics rounds questioned the motives of the house staff. The medical student said, "The residents don't like to be awakened at night. It's inconvenient and bothersome when a code is called." The student's attribution to the residents of self-interested motives may well have been false. The residents may have believed that the patient's quality of life was too poor to warrant resuscitation; or that he was simply too old; or that even if resuscitated, he wouldn't survive to walk out of the hospital. But whatever their actual beliefs may have been, the patient did not clearly or obviously fit the medical criteria outlined in the law for a DNR order.

The DNR law stipulates that at least one of the following conditions must exist before a DNR order may be written for an adult patient who lacks capacity:

1. The patient has a terminal condition.
2. The patient is permanently unconscious.
3. Resuscitation would be medically futile.
4. Resuscitation would impose an extraordinary burden on the patient in light of the patient's medical condition and the expected outcome of resuscitation.

The residents may have believed that resuscitation would be futile, in which case a DNR order would be justified. The medical student's description of the patient's condition, taken from the medical chart, did not suggest to a senior physician present at the conference that resuscitation would be futile. The patient was not diagnosed as having a terminal condition, nor was he unconscious.

The only remaining possibility lies in an interpretation of condition 4, a condition that requires a judgment call. It is easy to inject subjective values into a judgment that a particular treatment would impose an extraordinary burden on a patient. And there is no single, objective criterion for balancing that burden with the expected outcome of resuscitation for the patient. According to one view, young medical residents were stretching the notion of extraordinary burden to apply it to a case in which they viewed the value of a short extension of an 86-year-old man's life to be dubious. A more charitable view treats condition 4 as a threshold for allowing surrogates rather than physicians to make decisions about resuscitation for patients who have lost decisional capacity. On this latter view, the residents were respecting the patient's family in allowing them to make the final decision. The worrisome feature of this case, as presented at the conference, lay in the perception that the residents had convinced the family that the patient should be a DNR.

The judgments and actions of an attending physician are critical in the New York DNR law. In order to prevent the type of situation that occurred in the purple dot case—whereby physicians denied knowledge of or involvement in no-code orders, the orders were issued orally, and nurses who affixed the dots were the ones held accountable—the law clearly stipulates the role of the attending physician. In the case of patients who lack capacity, it is the attending physician (not the residents) who must find that one or more of the four medical indications exist for writing the DNR order. The attending physician must make the chart entry. In the case of this 86-year-old patient, the residents sought out another attending physician who was willing to make the chart entry. Although it is possible that this second physician independently arrived at the judgment that resuscitation would impose an extraordinary burden on the patient, participants at the ethics conference inferred, based on their own experiences, that an attending physician can always be found who will go along with the residents' treatment plan. This is one way in which the DNR law can be circumvented or misapplied.

Case 2: The patient was a 62-year-old man with a complicated medical history who now had end-stage heart disease. Unlike the foregoing case, in which a DNR order may have been inappropriately written, in this case a patient was badgered into having a DNR order reversed after he had initiated a request that the order be written. After having been on a ventilator and then removed, he told the nurses he didn't want ventilatory assistance again if he became short of breath. He asked to sign a DNR order, the physician in charge of his care carried out the proper documentation, and the patient signed the order himself. The man's family was at his bedside the entire time, and the nurses reported that the patient's brother yelled at him, saying that he should have the breathing tube.

Subsequently a second physician came by and reviewed the patient's chart, noting the DNR order. This doctor judged that the patient's problems might be reversible, that is, that he had a treatable condition. Therefore, he convinced the patient to reverse the DNR order. The patient went along with the physician's recommendation and with his family's urging that everything be done. Despite continued treatment, the man's condition worsened and he died 12 hours after the DNR order had been reversed.

When this case was presented at a combined medicine-nursing grand rounds, debate was intense. Some participants raised doubts about whether the patient really understood the meaning of a DNR order, what was entailed by an attempt at resuscitation, and the precise details of his own condition and prognosis. One resident questioned whether the patient was competent to decide, citing as reasons for doubt the level of oxygen in the patient's blood and his despair over his condition. The resident further suggested that the patient's request for a DNR order showed him to be suicidal and therefore "irrational." Others at the conference wondered what the patient's change of mind signified: did it mean he was genuinely ambivalent about his original request not to be resuscitated?

One additional fact was brought out as the discussion proceeded. The patient's personal physician (who was not directly responsible for his hospital care) had told the attending physician in the hospital: "Don't bring these papers [the DNR forms] to my patient. I don't want him to get depressed."

As is true of virtually all real-life experiences, this case was complex and contained several uncertainties that could not be fully resolved. One thing is clear, however: the ethical and legal obligation of the patient's personal physician to talk to his patient about resuscitation. Especially since the patient initiated the request for a DNR order, it was irresponsible for the physician to command others "not to bring the papers" to his patient. The law requires that when a patient requests information about DNR, a full discussion must be conducted. The law does not, of course, and should not prohibit a patient from changing his mind. The evidence brought out at the conference more than suggested that the patient was unduly influenced to alter his initial request, a request he made after having experienced what it was like to be on a ventilator. Someone commented that in the future, lawsuits will be brought against physicians for not following the clearly expressed wishes of patients who have decisional capacity, a sobering note on which the conference ended.

Case 3: A medical student was presenting a case at ethics rounds. In the course of his presentation, the student said, "And so they wrote a 'DNI order' in the patient's chart." I was puzzled by this term, which I hadn't heard before, and asked the student what that meant. He said, "That's a 'Do Not Intubate order.'" I turned to the senior physician who co-teaches these clinical ethics rounds with me. "Do you know about this order?" I asked him. The physician said he had not heard of such an order being written in a patient's chart, although discussions are frequently held about the advisability of intubating a patient or about the likelihood of eventually being able to wean the patient from the tube.

As the conference proceeded, it became clear that in this context, writing a DNI order was a creative new method devised by the residents to avoid the elaborate steps required for writing DNR orders under the newly enacted law. However, a DNI order can have other legitimate purposes. Since intubation is very often a component of full CPR, a decision not to intubate can be medically equivalent to a decision not to resuscitate. The residents' understanding, or perhaps misunderstanding, of the law led them to judge that only a DNR order legally requires them to hold a discussion with a patient who has capacity. By the simple device of creating a new "order," the house staff was seeking to revert to the common practices in place before a complicated, burdensome law was enacted. The surgeon who complained to me about the DNR law being "irrational and unnatural" added the comment that the law had spawned these DNI orders, "which came out of nowhere."

Even in the best circumstances, well-intended laws could not possibly cover all situations. A physician in charge of the residency training program in medicine at a teaching hospital put in writing the difficulties created by striving to comply with the New York State DNR law.

> From time to time I am asked to sign the DNR form to corroborate lack of capacity and appropriateness of the order in a particular patient. These occasions provide a fresh opportunity to discuss the general issue of DNR procedures with various house staff. They regularly raise the same issue.
>
> Invariably the problem occurs at 3 in the morning, not before or after. The patient is chronically demented, as determined by age, appearance, behavior, and assessment. There is never a family member in scenario 1 (scenario 2 to follow). There is evidence of wasting, and it is quickly determined on examination that the patient is suffering from widely metastasized cancer or some other equally fatal disorder. The attending physician will be in to see this patient sometime the next day. Everyone on the team agrees that DNR would be burdensome to the patient and "futile" in the normal sense of that word. Without a DNR order from the attending physician, the house officers cannot write an order. If that is so, once the patient is in bed and under the care of the nurses, a cardiac arrest will result in the nurses calling a "code." If the informed house officers respond, the curtain is pulled and nothing meaningful is done. If a "strange" team responds or the anesthesiologist arrives first, the patient is intubated and CPR initiated. Why, the residents ask, should the patient be abused? Why are they required to do something that they believe to be medically and ethically inappropriate?
>
> In the second scenario there is a family urging that nothing more be done to the unfortunate patient, who has suffered enough. The informed team may use this to override the nonexistence of a signed DNR order. However, the nurses are not inclined to refrain from calling a code without a written DNR order, and that requires two attending physicians in a patient who lacks capacity at 3 A.M. If a code is called by the nurses, the responders may be uninformed and initiate CPR. The only way to avoid that is for a label to be used, and we are back to purple dots.
>
> How can we help the house officers faced with these decisions?

In response to this question, a physician well acquainted with New York's DNR law asserted: "Set up a rational policy!" He observed that the law does not force this series of events, a reminder that laws require sound policies to be developed by institutions for the purpose of implementation.

The predicaments described in this physician's memo could be resolved, or at least made less problematic, by revising some of the procedures. Even if these difficulties cannot be altogether eliminated, the doctor's attention to them represents a concerted effort to comply with the law and, at the same time, to do the right thing ethically and medically. However, in the hands of unscrupulous physicians, the law can be used for ethically questionable and even malevolent purposes. The following case was unprecedented in the experience of a colleague who reported it at a case conference.

The physician who presented the case was a senior neonatologist, called by a junior colleague who was working at one of the affiliated hospitals of the academic medical center. The junior colleague, Dr. McCord, had been asked to attend at a forthcoming cesarean section, as is common practice. When Dr. McCord met the obstetrician who was to perform the cesarean section, he was informed that the fetus was "dysmorphic"—physically abnormal, with a number of anatomical deformities. The condition was revealed by an ultrasound examination that had been done the previous day. The reason for the cesarean section was that the baby's head was too large for the mother's pelvis. There was no family history of this or other birth anomalies.

The parents were very upset by the ultrasound scan of a physically abnormal baby. They did not want to have a dysmorphic child. They insisted that the only acceptable outcome was the death of this baby. The mother was also unwilling to give up the baby for adoption. The solution proposed by the obstetrician was to have the parents sign a DNR order for their unborn child. Since babies often need help in breathing at birth, the intent was to not assist the baby's breathing at the time of delivery in the hope that the infant would die.

Dr. McCord was shocked by this unprecedented use of a DNR order in an unborn child. He was dismayed that the obstetrician in charge of the case could be so influenced by the parents' disappointment that he would try to ensure the infant's death from failure to attempt resuscitation. Uncertain about whether there was anything he could do, Dr. McCord telephoned his senior colleague for an emergency consultation.

The consultant told Dr. McCord that obtaining a DNR order for an unborn infant was both illegal and stupid. It was illegal because it was impossible to know in advance whether the infant would meet the medical criteria spelled out in the law for writing a DNR order for a child. If any of the medical indications for writing a DNR order are present, then two licensed physicians must attest to the presence of at least one of these conditions before the infant's parents may be asked if they wish the order to be written.

The action of the obstetrician was stupid because it could lead to an even worse outcome than the death of the infant. If the baby could breathe on its own, yet was asphyxiated, then the most likely outcome would be survival, but with moderate to severe brain damage.

The consulting neonatologist advised his junior colleague that if the baby did not have a fatal illness and was not dying at the time of birth, active resuscitation should be performed. When Dr. McCord informed the obstetrician of this recommendation, the obstetrician replied, as he ripped up the DNR order: "Okay, now it's in your hands. You deal with the family." Dr. McCord informed the father that if the infant was viable, he would have to resuscitate it. In fact, the baby did require resuscitation and needed to have a breathing tube inserted.

Participants at the conference felt sympathy for these parents, who were so deeply disappointed and despondent at the birth of a less than perfect baby. But opinion was unanimous in support of the neonatologist's view that a prenatal order not to resuscitate was illegal and stupid, as well as unethical. One physician contributed the information that "fetal DNR orders" have become routine in some hospitals, and that this maneuver was being suggested by hospital lawyers. The lawyers have apparently been advising their clients that they are on safer legal ground if they establish a practice of writing fetal DNR orders than if they fail to treat a handicapped newborn aggressively in the absence of a DNR order.

That advice is almost certainly flawed from a legal point of view, and from an ethical perspective it is obtuse. A decision not to resuscitate an infant can be ethically justified, but only when the infant's actual condition at birth can be properly evaluated and its prognosis determined with a reasonable degree of accuracy. The lawyers who issue advice to write fetal DNR orders do so in the belief that they are protecting their clients from liability. But if physicians act on such advice, they are failing in their medical and ethical responsibilities.

Is Law a Good Solution to Ethical Problems?

The episodes and practices that have occurred in response to the recently enacted DNR law in New York State give rise to the question: Is legislation the best solution to ethical problems? Put in such a general way, that question is impossible to answer, except by saying: Sometimes it's not. It is better to inquire whether, on the whole, a particular law proves to be good or bad, and in what specific respects. Not every situation in which a solution creates problems would have been better off without the solution altogether. Although it may be true in general that once laws are enacted they are hard to change, it is also true that laws are repealed or amended with some frequency. While none of the major problems cited earlier with regard to New York's DNR statute have been altered by amending the legislation, some useful modifications in the public health law have already been enacted.

These changes deal with improving procedural requirements for issuing DNR orders. Improvements include allowing surrogates for patients to give oral consent in lieu of the current requirement that consent to a DNR order be issued in writing. It has been observed that family members—especially parents of dying children—view having to sign a form as signing a death warrant, so the amendment allows for a surrogate to consent orally to two adults, one of whom must be a physician affiliated with the hospital. Another procedural change extends the time required by a physician to review the chart of a patient with a DNR order. The earlier three-day requirement for review was found to be too rigid in practice, so the review period has been lengthened to seven days. Other rules relate to the validity of DNR orders when a patient has been transferred from one facility to another, and to alterations in this law now that New York state has passed health care proxy legislation.

A Department of Health Memorandum noted that studies have shown that some of the procedural requirements in the original law "are not reasonably necessary for the protection of patient interests."[27] The Memorandum concluded by stating:

> The DNR law was a unique effort to clarify and balance the rights and obligations of patients, their families and health care professionals in difficult decisions about resuscitation. As might be expected given the complexity of the issue and the novelty of the approach, experience has demonstrated areas on which the law may be improved. This bill will refine and improve the law and reduce some of the sources of tension that have been noted in practice, while retaining the central ethical premises it embodies and protections it affords.[28]

Amending procedures in the law will surely not preclude attempts to circumvent it by charades such as Hollywood codes or inventions of new categories such as DNI orders. Despite the comments cited above from disgruntled or beleaguered physicians, it does seem that the DNR law accomplished one main objective: promoting patients' and families' participation in DNR decisions. At the very least, it has succeeded in enhancing patients' rights to informed consent and refusal. No doubt, the law has also heightened physicians' awareness of the importance of adhering to objective medical criteria before deeming a patient a candidate for DNR orders.

This law was designed to combat flagrant abuses of patients' rights, and in that regard it has brought a noticeable improvement in the situation that previously prevailed when covert practices were the rule. However, it may best be seen as an interim measure until a more comprehensive law is enacted that would address criteria for withholding or withdrawing other life-sustaining treatments for incapacitated patients who have not previously expressed their wishes about aggressive therapy. But there is no doubt about one thing: passage of a law on orders not to resuscitate has clarified matters for hospital administrators and legal departments.

Uncertainty is likely to remain over whether efforts to enact laws or institute regulations designed to protect patients do more good than harm on the whole. The answer is likely to vary in different circumstances, depending on whether a particular law is clear or too complicated to understand; whether its implementation is straightforward or cumbersome; and whether those whom it governs succeed in finding creative ways to circumvent it, among other factors.

Enacting laws cannot be a solution to all ethical problems, whether they occur in medical practice or elsewhere. Dilemmas are bound to arise because of conflicts of moral obligation, uncertainty about which value should take precedence when clashes occur, and different assessments of the rights and interests of patients. But where these were matters once largely resolved within the physician–patient relationship, today they involve lawyers and risk managers, who occupy a prominent position in hospitals throughout the country. That position permits them to intrude in the doctor–patient relationship, often to the detriment of physicians and patients alike. What is the role of these bedside bureaucrats, and in what ways do they intervene into the rightful prerogatives of doctors and patients?

Notes

1. For a brief, illuminating history of resuscitation, see President's Commission for the Study of Ethical Problems in Medicine and Biomedical and Behavioral Research, *Deciding to Forego Life-Sustaining Treatment* (Washington, D.C.: U.S. Government Printing Office, 1983), pp. 231–34.

2. Ibid., p. 234.

3. See, for example, Mitchell T. Rabkin, Gerald Gillerman, and Nancy R. Rice, "Orders Not to Resuscitate," *New England Journal of Medicine*, vol. 295 (1976), pp. 364–66; and "Optimum Care for Hopelessly Ill Patients: A Report of the Critical Care Committee of the Massachusetts General Hospital," *New England Journal of Medicine*, vol. 295 (1976), pp. 362–64.

4. Susanna E. Bedell and Thomas L. Delbanco, "Choices About Cardiopulmonary Resuscitation in the Hospital: When Do Physicians Talk with Patients?" *New England Journal of Medicine*, vol. 310 (1984), pp. 1089–1093.

5. Ibid., p. 1089.

6. *Deciding to Forego Life-Sustaining Treatment*, p. 239.

7. Ibid., n. 31.

8. David Margolick, "Hospital Is Investigated on Life-Support Policy," *New York Times* (June 20, 1982), p. A34.

9. Ronald Sullivan, "Hospital's Data Faulted in Care of Terminally Ill," *New York Times* (March 21, 1984), p. B-1.

10. Report of the Special January Third Additional 1983 Grand Jury Concerning "Do Not Resuscitate" Procedures at a Certain Hospital in Queens County (hereinafter, Grand Jury Report), State of New York, Deputy Attorney General For Medicaid Fraud Control, dated Queens, New York, February 8, 1984, p. 3.

11. Ibid., p. 4.

12. Ibid., p. 12.

13. Ibid., p. 13.

14. Ibid., p. 15.

15. Ibid., p. 16.

16. Ibid., pp. 13–14.

17. *Deciding to Forego Life-Sustaining Treatment*, pp. 248–49.

18. Grand Jury Report, p. 19.

19. Ibid., p. 23.

20. New York Public Health Law Article 29-B, Orders Not to Resuscitate, Section 2961, Definition 9.

21. Robert N. Swidler, "The Presumption of Consent in New York State's Do-Not-Resuscitate Law," *New York State Journal of Medicine*, vol. 69 (1989), pp. 69–72.

22. Ibid., p. 69.

23. Ibid.

24. Ibid., p. 72.

25. All quotations are from *The New York Doctor* (June 27, 1988), p. 8.

26. Not the actual name.

27. Peter J. Millock, General Counsel, DOH No. 23R-91, Memorandum, p. 3.

28. Ibid., p. 6.

3

Bureaucrats at the Bedside:
Risk Management

Anyone familiar with the workings of hospitals today can recount tales of administrators who insert themselves between patients and their physicians or thwart family members as surrogate decision makers. The following two cases are illustrative.

Marjorie Brown (not her real name) was a 62-year-old schoolteacher with amyotrophic lateral sclerosis (ALS) who had been an inpatient for 4 years, dating from the time her disease had progressed to the point where she could no longer take care of herself at home. She was unmarried and had previously lived alone in an apartment in a two-family dwelling that she owned. Her sister lived in the other apartment with her own family. Her brother resided elsewhere in the city.

At the time her case was brought to the ethics committee, Marjorie Brown was completely paralyzed except for the ability to blink her eyes. She was on a ventilator and had a feeding gastrostomy. In addition to the hospital nursing staff, the patient was attended by a private duty nurse. That nurse read to her, groomed her, and conveyed most of her wishes to the other caregivers in the hospital. Ms. Brown communicated with the nurse by means of a hand-lettered chart containing the letters of the alphabet. As the nurse pointed to each letter in succession, the patient blinked once to respond "yes" and twice to respond "no," spelling out words and forming sentences in this painfully slow fashion.

Even before her condition deteriorated to its present stage, Ms. Brown had begun to express the wish to end her life. Her sister visited regularly but not very frequently, and her brother came to the hospital only occasionally. Ms. Brown was alert and oriented, and her communications, although labored, revealed her to be fully lucid. At the time she first expressed her wish to have the ventilator turned off, a psychiatric consultation was sought. At that time, and on several subsequent occasions, the patient was judged to be mentally competent and not to be suffering from clinical depression.

As Marjorie Brown's requests for termination of life support became more frequent, the attending physicians, who were prepared to respect her wishes, decided to consult the hospital ethics committee. They informed the patient of that decision, and Ms. Brown consented to the consultation.

In May, a consulting team from the ethics committee, consisting of two physicians, a nurse, and an ethicist, met first with the entire staff on the unit. The assembled group included the attending physicians who had known the patient during her entire four-year stay, members of the nursing staff, a social worker, the patient's private duty nurse, and respiratory therapists on the unit. The medical and psychosocial aspects of the case were presented in detail, and the consulting team asked some questions and discussed the pertinent ethical and legal issues with the staff. The meeting lasted for about an hour. At the end of the meeting, we asked whether anyone in the assembled group objected to honoring the patient's request to remove life support or had lingering reservations. No one spoke up, and most people shook their heads "no."

We were told, however, that there might be some opposition by the family. The staff thought that Ms. Brown's sister had a self-interested reason for keeping the patient alive: she feared for her security and housing needs, as she lived in the house owned by Ms. Brown. The staff also thought that the brother might oppose withdrawing treatment, stemming perhaps from his own guilt at not having been more attentive to his sister during her long illness and hospitalization.

Next, the consulting team went to the bedside. We introduced ourselves to Ms. Brown and asked her permission to pose some questions. She blinked once in reply. We asked what her wishes were; how long she had felt this way; why she wanted to terminate life-sustaining treatment; and whether she wished to consult a member of the clergy, family members, or anyone else. While her nurse held the card with the letters of the alphabet, the patient gave her answers in full sentences that she formulated. The consulting team had decided that it was important not to pose simple yes-or-no questions, but to allow the patient to express her thoughts as completely as possible in her own words. She indicated that she wanted to be asleep when the ventilator was turned off.

When we had asked all the questions we thought appropriate, we informed the patient that we would report back to the ethics committee and try to do everything possible to have her request granted. We then asked Ms. Brown if she had any questions she wanted to ask us. With her nurse holding the alphabet chart, the patient spelled out the letters W–H–E–N. We had to reply that it might take some time before her request could be fulfilled, but we would try as hard as we could to move the process along. Her eyes expressed disappointment at this reply.

The physician who headed the consulting team wrote a lengthy note in the patient's chart documenting the consultation. At the next meeting of the committee, we reported on the meeting with the staff and the visit to the patient. A long discussion ensued, with various members of the committee stating that this case represented one in which the patient is clearly competent, has clearly thought through her decision, has had four years of experience with the ventilatory support she wished to have withdrawn, and, therefore, is completely informed about the nature of the treatment she is refusing. The committee spent some time discussing the situation involving the patient's sister and brother and their potential opposition. It was agreed that from an ethical perspective, their wishes could not override those of the patient. Yet we recognized that the hospital's risk management personnel might have some worries about acting in opposition to a family's demands.

With no dissenting voices, the committee concluded that the patient's wish should be honored as soon as possible. The physician who had headed the consulting team was authorized to write a follow-up note in the patient's chart documenting the full committee's conclusion. The case was then turned over to the hospital administration, and for the time being was out of the committee's hands.

Following that month's committee meeting and placement of the consulting team's note in the patient's chart, materials were sent to the corporate governing body of the hospital (this hospital is part of a larger hospital system). The purpose was to obtain legal consultation and final authorization to remove the ventilator. The legal affairs division of the corporate governing body informally communicated to the hospital administration that it would be permissible to withdraw respiratory support from the patient without obtaining a court order. But at that point, formal written notification and procedural details had not yet been provided to the hospital administration.

At its next monthly meeting, the hospital's risk manager brought the committee up-to-date on these developments. Committee members lamented the delay in taking action to follow the patient's directive. One person observed that at this point, the hospital was providing treatment against the patient's clearly expressed wishes. Others agreed that to persist in giving treatment against the clearly expressed wishes of this patient was ethically unconscionable. Someone surmised that the hospital administration might be fearful of a reaction in the community to the act of shutting off the patient's ventilatory support, even if the ethical and legal questions were resolved in favor of the patient.

Discussion then turned to some practical issues. Who, in fact, would be responsible for administering adequate sedation and turning off the patient's ventilatory support? It was agreed that the appropriate clinician was the attending physician on the ward service. It was also agreed that any particular clinician had the right to refuse to carry out such an action if it was felt to be inconsistent with his or her personal beliefs or professional obligations. If that were to happen, then the responsibility for carrying out the patient's wishes would follow the supervisory hierarchy of the hospital, with the medical director being ultimately responsible. The consulting team reminded the committee that none of the clinicians with whom they met during the original consultation expressed any disagreement or hesitation regarding the obligation to honor the patient's request.

The committee recommended that the liaison psychiatry service pursue this issue, with a plan to provide support for the clinicians who would be responsible for turning off the ventilator. The director of liaison psychiatry was a member of the ethics committee and endorsed that idea. Concern was then raised that holding a discussion with the ward attending physician or any other clinician might be premature in view of the fact that no formal permission had yet been received from the hospital's governing body. In reply, it was noted that insofar as the staff had already contemplated such an action with great seriousness, a need existed to discuss the clinicians' reactions to what would be done and to counsel them accordingly. It was generally agreed that the sooner such counseling and preparation began, the more effective and useful they would be. Also discussed was the need to assist clinicians in dealing with the patient's family.

The meeting ended with an agreement that the hospital administration should be strongly encouraged to develop whatever practical procedures would be required for the clinicians to accede to the patient's request. The committee decided to draft a letter to the administration reiterating the ethical analysis that had been provided earlier.

By the time of the next monthly meeting, the committee chairman had received word that formal written permission to honor the patient's request had been granted by the hospital's corporate governing body. However, the hospital's local administration continued to stall. Shortly after the committee met, the hospital administration convened another meeting that included the patient's clinicians, the medical director of the hospital, liaison psychiatry, a representative from the ethics committee's consultation team, and the hospital's risk manager. The risk manager expressed assent to the unanimous urging of the group to implement patient's request as soon as possible. The next day, Ms. Brown's physician again spoke to her, this time with her sister present. The patient reaffirmed her wish to have the respirator removed, and her sister did not indicate any opposition. Ms. Brown was given valium, followed by morphine. As she became sleepy, her private duty nurse continued to sing to her. Then the respirator setting was changed. Three to five minutes later, Marjorie Brown died.

When an adult patient with decisional capacity requests the termination of life support, the ethical issues are largely *procedural*. Concerns focus on determining the patient's capacity to make the decision, ensuring that the wish to die does not stem from a reversible depression, when and by whom life-sustaining measures will be withheld or withdrawn, and other steps in the process. As the case of Marjorie Brown demonstrates, these concerns, as well as other procedural matters, were taken up by the ethics committee as part of its role in dealing with the request by the medical and nursing staff caring for the patient.

At a subsequent meeting of the ethics committee, it was reported that some of the nurses had had a negative reaction following the patient's death. One nurses' aide became very upset and had to go home. The committee discussed these developments, concluding that the negative reactions of members of the nursing staff could be viewed as normal grief responses rather than as an indication that something ethically wrong had occurred. It was also noted that some nurses on the ward had asked not to participate in the process. A physician observed that the emotions raised by this case are different from those normally expected in the practice of medicine. The committee concluded its retrospective discussion with the observation that it was worthwhile and even necessary to devote so much time and effort to this case. There would surely be more cases raising similar issues in the future. Although future cases might go more smoothly from a procedural perspective, they are not likely to be any easier emotionally on the staff and committee members who become involved.

There are still people who have ethical qualms, usually stemming from their religious convictions, about removing life support from patients—even patients who are alert, have decisional capacity, and have experienced the burdens of the medical treatment they seek to have withdrawn. But the case of Marjorie Brown did not involve any such opposition on ethical or religious grounds. A hospital administrator, for reasons that were never made clear, acted for months to delay recognition of the patient's moral and legal right to refuse continued medical treatment, causing her considerable distress in addition to her already serious medical condition. In the end, he did not succeed in thwarting the patient's wish. One can only speculate about whether the patient's request would have been respected at all if the ethics committee had not acted insistently to advocate for her wishes.

A second case, which occurred in a different hospital, involved a patient with AIDS who was in the final throes of his illness and in extreme suffering as he deteriorated, and who, like the patient with ALS, requested removal of his ventilator. His physicians, members of the ethics consultation team, and other caregivers concurred. The patient requested that he be sedated with morphine so that he would not be awake and alert when the artificial respiration was withdrawn. His physician was prepared to administer morphine to the patient and to remain at the bedside throughout the process to ensure the patient's peaceful passage from life to death. The doctor perceived his dual obligation to respect his patient's right to have medical treatment withdrawn, and at the same time to relieve the patient's suffering.

Having been alerted to the situation, the hospital's chief executive officer (CEO) intervened. He expressed a fear that the cause of death might then be the morphine—which suppresses respiration—instead of removal of the life support. In that case, the CEO reasoned, doctors would be responsible for "killing" the patient, and the possibility of an indictment for murder could not be ruled out. The administrator insisted that a court order be sought. The medical staff and the ethics consultation team were worried that despite the clear legal right of competent patients to refuse medical treatment, a trial court judge in the hospital's jurisdiction might refuse to honor the patient's request. The jurisdiction in which the hospital is located is known to have judges whose personal convictions about preserving life supersede their willingness to recognize legal precedents or ethical principles. The case was eventually resolved by an alliance between the physician and the patient, who together agreed that the patient would not be given the morphine in advance but that the physician would wait until manifest suffering was evident. This succeeded in convincing the CEO that the doctor would then be acting in his medical role of relieving his patient's suffering rather than as a euthanizer administering a lethal dose of morphine. Ethics consultants were left wondering whether the way the case was resolved violated the principle of beneficence. Although the patient's autonomy was ultimately respected, it was done at the cost of ensuring that manifest suffering had begun before the physician was permitted to administer morphine.

The newly created role in hospitals of "risk manager" marks the official recognition of a role that has become central to the way hospitals are run. Their role is clearly defined: "to render advice to and initiate action for an employer institution or agency designed to avoid or minimize potential legal, and hence, financial loss for the health care provider."[1] Although they may fulfill several different administrative functions, risk managers regularly act as enemies of patients. People may disagree over whether the many and detailed maneuvers to minimize liability are warranted, and thus whether the activities and even the presence of these administrators can be justified. I contend that current risk management practices in hospitals have risen to a pitch of near hysteria. They embody actions that are unprecedented in their intrusiveness into the doctor–patient relationship and are unethical in violating the rights of patients.

Although my indictment sounds extreme, I am not alone among scholars in the field of bioethics in this assessment. The lawyer-ethicist George Annas has written:

The quest for "100 percent immunity" is both unrealistic and unprofessional, and evidences a desire to put one's self-interest above the interests of individual patients. So pervasive is the desire for self-protection that in a number of instances even brain-dead corpses have been brought to court for judicial permission to cease "treatment." . . . Physicians should know at least enough law to be able to tell when the advice their lawyers are giving them is so incredible that it is most likely wrong. . . . [P]hysicians should realize that there are no 100 percent guarantees in law any more than in life, and that part of being a professional is taking responsibility for decisions within one's professional competence.[2]

A related criticism is that risk managers misinterpret and misapply laws in a way that not only works against the patient's and family's ethical interests but also restricts physicians' ability to balance and respect conflicting ethical principles.[3]

It is evident to anyone who overhears doctors discussing their work that fear of legal liability drives much of their behavior. Hospital attorneys and administrators exhibit an even more pervasive fear, which has now become institutionalized. Annas's recommendation that physicians be more circumspect in taking advice from lawyers can hardly be followed in the typical hospital setting today. Physicians are loath to act independently when they perceive any prospect, however remote, of liability, and so in such cases, before they do anything, they "call risk management." It is the rare physician who acts deliberately against the decision of a risk manager.

Based on evidence regarding the real risks of liability, drawn from situations similar to those in which risk managers act to protect the institution, the typical responses of risk managers are wildly overreactive. If people in ordinary life were to act in accordance with the minuscule probabilities on which risk management bases its decisions, we would all be in a constant state of paralysis. As one writer quipped: "the risk manager effectively acts more as a paid paranoid than as an enabler of appropriate clinical and ethical health care delivery."[4]

Physicians have become so intimidated by risk management that they fail to act in ways they sincerely believe to be ethically responsible. A tragic episode, which may well have involved actual malpractice, involved a child who was brought from a developing country to a major medical center in the United States for surgery. Failure to monitor the child properly after surgery resulted in complications followed by a deterioration of his condition. He was left with irreversible massive brain damage. Much debate ensued among the physicians about whether to tell the truth to the family about the causes of their son's condition. To this point they had done nothing, hoping that the boy's poor condition could be reversed. Physicians had dodged the opportunity to discuss things with the family, and the boy's parents had failed to ask direct questions. A majority of those present at an ethics case conference believed that the morally correct action was to disclose the facts to the family.

But an overpowering circumstance prevented them from doing so. Risk management would have to be consulted. Why? The hospital administration had initially learned of the case through an incident report filed according to proper procedures by the physicians after the untoward event. Now, the physicians contended, they could not go ahead and do anything further by way of disclosure to the family without consulting risk management. Based on past experience, they were certain that the hospital would not allow the patient's family to be told what had actually transpired. These physicians were probably correct in their assessment of what this particular hospital administration would do. But other experienced physicians observe that not all risk managers would cover up an event like this. In fact, one doctor surmised, when he learned of the case, that most administrators *would* disclose.

In any event, if the treatment of the child did actually involve malpractice, there was a real risk that a legal action brought by the parents would be successful. Whether the family would actually initiate such an action is a matter of speculation, since they were foreigners, not prone to the litigious behavior so common in the United States.

Whether physicians have an obligation to disclose medical error, and whether patients have a correlative right to be told the truth about untoward events, can be subjected to an ethical analysis independent of the risks of a malpractice suit. According to one position, doctors owe patients (or, in cases like this, the family) an explanation of ongoing aspects of treatment as part of the larger obligation to communicate the risks, benefits, and alternatives. This ethical position focuses on the rights and duties inherent in the physician–patient relationship. One of the fiduciary duties of physicians is to be truthful to their patients, placing patients' interests above other concerns that may militate against disclosure.

A different ethical perspective focuses on the consequences of the actions open to physicians, including the consequences to the family of being truthful about the causes of their child's present condition. This analysis asks whether the good consequences of disclosure outweigh the bad or vice versa. The family might be made more unhappy at the news that the cause of their child's poor condition was preventable. They might lose faith in the medical profession, and fail to trust doctors and hospitals in the future. In a consequentialist analysis, it is fair to bring in the potential negative consequences to the doctors and hospital that would emanate from a malpractice suit being brought and possibly won by the parents of the child.

Of course, an opposite set of consequences could also be envisaged: the parents might be relieved to be given some explanation rather than remain in ignorance and doubt; they might be grateful to the doctors for being forthcoming and respect them for admitting a medical error; and it might not even occur to them to initiate a lawsuit. To conduct an ethical analysis by projecting the good and bad consequences of actions is a respectable methodology. But the analysis is only as sound as the ability to predict the likelihood of the alternative courses of action. The standard procedure of risk management is to project the worst possible consequences, however remote they may be, and direct hospital personnel accordingly.

It is true that malpractice suits have increased and that the size of awards has grown. But the response by hospitals has been unrealistic and unreasonable: to try to eliminate all conceivable risks of liability. Whether this no-risk policy is directed at eliminating civil or criminal liability, it has not only disenfranchised patients; it has also intruded into the doctor–patient relationship even in situations where physicians and patients are in complete agreement about a course of treatment.

Who are these risk managers, and what is the origin of their role in the hospital? Some risk managers have law degrees, but most do not. Some are nurses who rose to the rank of supervisor and then moved into hospital administration, often after obtaining a master's degree. Others come from the ranks of hospital administrators, some with a degree in hospital or business administration. More rare are individuals with an advanced degree in a field such as sociology, and still others made their career in health planning or administration and were around long before the occupation of risk manager was invented. Large medical centers typically have an office of risk management in addition to in-house counsel. The staff of lawyers works together with risk management both in devising hospital policies that affect patients and in dealing with individual cases in which anyone suspects that there may be a risk of some sort.

The overall movement can be traced back to the late 1960s and early 1970s, when efforts were begun in industrial and other workplaces to reduce the costs of liability payments by underwriters and insurance companies. The trend widened, and in the 1980s risk management offices began to be established in hospitals. The situation can best be described as a growing tendency to assign corporate responsibility for individual actions. Although it was once common practice to sue only the doctor involved in a case, now it is both doctors and hospitals who are the targets of lawsuits.

The office of risk management was originally created to deal with concerns about possible liability arising out of incident reports in the hospital: a patient falling out of bed, a visitor slipping in a puddle of water in the corridor, an inadvertent injury to a patient in the course of treatment. These concerns focused on the possibility that a patient or family member might sue the hospital, and efforts were bent on minimizing any such liability that could result from actual or foreseeable untoward incidents. It is a legitimate function of a risk management department to devote its efforts to raising the standards of safety and medical practice in order to reduce the underlying causes of untoward events and ensuring lawsuits. More often than not, however, their strategic energy has focused on "damage control."

The original worries about legal liability have now expanded to encompass everything that might place the hospital in a bad light. Risk managers are now charged with the task of minimizing risks other than those of liability. They look out for the projected risks of bad publicity, the actions of a disgruntled employee, or the possible political ramifications of a medical decision or hospital policy. A nurse who disagrees with a consensual decision made by the patient and doctor to withdraw treatment might "blow the whistle," and headlines in the next morning's tabloid newspaper would cause trouble for the hospital. Concerns are voiced about the possibility of a negative reaction in the surrounding community, so risk management gets involved in the case. Even when the patient has no family, and there is no one around who would sue the hospital, risk management is brought into the case. One of the peculiar features of this situation is the nearly automatic response by many physicians to call risk management whenever the slightest uncertainty is voiced about an ethical matter or vaguely perceived to have legal implications.

In one instance, a hospital's bioethics committee was convened for a case consultation to discuss whether aggressive treatment should be discontinued for an infant with a terminal illness. As is the customary practice, the parents were invited to meet with the committee. It is also the practice of this committee to invite the caregivers who have been involved in the baby's care and treatment in the hospital. For this meeting, one of the pediatricians had prepared a detailed written summary of the infant's course from birth to the present. The baby's father was present at the beginning of the meeting to express his and his wife's views about continued treatment and to ask the committee any questions he wished. A copy of the summary was given to him when it was distributed to the committee members. The risk manager joined the meeting after it had begun. After about a half hour, the father left the committee meeting and the risk manager learned that a copy of the written summary of the case had been given to him. The risk manager was furious. "Why was he given a written copy of the summary? That wasn't for him. He shouldn't have been given a copy of the summary!" When I asked "Why not?" the risk manager replied defensively: "It's our policy not to give written summaries to patients or families." It is fair to conclude that she did not have a reason she could articulate.

The chairman of the committee assured the risk manager that there was nothing damaging or incriminating in the summary prepared by the pediatrician. The risk manager was not at all interested in that fact, but kept reiterating the importance of never giving patients or families written copies of case summaries prepared by medical staff. There *might* be something in a written summary, some time, in some case, for some patient, that could be used in a lawsuit brought against the hospital. Alternatively, if the document fell into the wrong hands, there would be those much-feared headlines in the next day's newspaper. As a result, the general practice in this hospital, from a risk management perspective, is simply never to give a written summary to a patient or family member.

My experience with risk managers in hospitals has largely confirmed the judgment that they act as paid paranoids. A physician colleague recently admonished me to soften my view, remarking that "They're not really bad people." I do not for a moment think that risk managers are "bad people." They are doing their job and trying to do it well. One risk manager confided to me that he is often deeply troubled about that role. "I go home at night, after a day's work, and worry about the ethics of my job. It often doesn't come out best for the patient." This administrator can be commended from an ethical standpoint for his self-conscious approach to his job, despite the fact that his role routinely requires him to place the hospital's interests above those of the patients.

It is true, of course, that there are good and bad risk managers. The best risk manager I have known is someone who had been involved at the higher ranks of hospital administration long before the term "risk manager" became common parlance. As a vice-president of a large urban medical center, he spent many years looking out for the hospital's interest in minimizing liability. As the role of risk management became official, he assumed the new title as well as the increased duties. He cared a great deal about ethics, and in fact was the moving force behind the establishment of the hospital's ethics committee during the 1970s. I suspect he was not unique in his belief that making an effort to do what is ethically right and appropriate comports with good risk management. He retired some years ago, but his philosophical pronouncement about his role still lingers: "The job of a risk manager is knowing what risks to take."

Administrative Procedure: A Sample

An example of the role of risk management in one hospital's procedures is spelled out below in an account adapted from its administrative policy. The cases covered are those involving withdrawal of treatment from patients who are not capable of giving informed consent, in which the physician agrees with the plan. I believe it is important to have some mechanism to protect vulnerable patients from actions that may be against their best interest. My point in offering this illustration is not to question the need for procedural safeguards, but rather to demonstrate the entrenched administrative role of risk management in the modern hospital.

The policy describes a four-level review process. Level I is entitled "Risk Management and Physician." At this stage, risk management "will review the case with the responsible physician." It is up to risk management (rather than the physician, patients' family, or others) to determine whether the case can be resolved at this point. If risk management determines that it cannot be resolved, it goes to Level II review.

Level II is entitled "Risk Management, Physician, Caregivers, Patient, Family, and Friends." Risk management initiates an expansion of the review among these individuals in order to gather relevant information. Following this review, it is once again the job of risk management to determine whether or not the issue can be resolved. If not, it goes to Level III.

This level is entitled "Risk Management, Caregivers, Patient, Family, Friends, Legal Affairs, Bioethics Consultant, and Medical Director." The two possible outcomes of this review are that (1) the refusal of treatment will be honored and risk management will document in the patient's medical record the basis for refusal or (2) further administrative review is required.

If (2) is the outcome, Level IV is initiated: "Review by Senior Administration." In preparation for this level, risk management is to request signed and notarized affidavits from the patient's family and/or friends regarding any prior wishes expressed by the patient about his or her future medical care. These affidavits are then reviewed, along with any other relevant information, with the department of legal affairs. That department prepares a written opinion concerning the case.

Risk management then contacts the hospital's ethics committee (chairman or designee). A new group then convenes, including members of the medical and nursing staff, hospital administration, psychiatry, department of legal affairs, a bioethics consultant, and the medical director of the hospital. This group then conducts a thorough review of the case and issues its opinion. With this report in hand, the case is then reviewed once again by risk management and legal affairs. If it is deemed appropriate for treatment to be withheld or withdrawn, proper steps are taken to implement this decision. If it is decided that treatment cannot be withheld or withdrawn without seeking a court ruling, the next decision is whether or not to seek that ruling. Risk management is to notify the attending physician and the family of the decision.

The final steps are carried out by risk management: review of the documentation in the patient's record; involving the patient's family and relevant medical and nursing staff in any ensuing discussion; and keeping these parties informed of any further decisions regarding withholding or withdrawing treatment.

If the aim is to ensure that medical treatments are not inappropriately withheld or withdrawn from patients, this process is highly likely to guarantee that result. It is hardly a model of efficiency, but the value of efficiency is ethically subordinate to that of protecting patients from being harmed or wronged. On what basis, then, can these elaborate procedures be faulted? Two different answers come to mind.

The first is the notable absence of substantive criteria that could justify a decision either to terminate treatment or to reject that option. The policy is an exquisite example of the role that process has come to play in many institutions. If we have a proper process, this theme resounds, then we don't have to worry about substance. Elaborate procedures have become a substitute for sound, substantive criteria for action. It is true that attention to procedures is often necessary to ensure that fairness and rights of due process are respected. But they cannot entirely supplant the need for criteria for judging actions to be right or wrong.

The second basis for questioning the structure set up in the policy is the central role played by risk management. What expertise do these managers bring to the proceedings and deliberations into which they intrude? Unlike the patient or family, the physicians, nurses, and other caregivers, the department of legal affairs, the ethics committee, the bioethics consultant, and even the senior administration of the hospital, the risk manager has no role other than that created by the perceived need to manage risks in hospitals. It is a function that ensures that hospital bureaucrats play a part in decision making at the bedside.

Often the issue is not one of different and conflicting perceptions of risks, but rather one of *control*. The result is that risk management ends up controlling patients, their families, and physicians. Hospital administrators have refused to permit competent adult patients to reject burdensome treatment even when physicians concur with the patient's wish. It is not uncommon for administrators to request that a court order be obtained whenever there is a shred of doubt (which almost always exists) about what the law says. Where it was once physicians who overtreated patients because they believed it was their moral obligation to continue therapy, it is now hospital administrators and risk managers who more often insist on overtreatment out of fear of medical-legal liability. It is not a great exaggeration to view risk managers as enemies of patients.

The practice of bringing cases to court prospectively, seeking judicial approval for a contemplated action, is a disturbing trend. A directive by one hospital administration stated that in all cases where a decision to withhold or withdraw treatment from an incapacitated patient is involved, the hospital must go to court to request permission. Even when families and physicians are in complete agreement, the directive stated that a court order must be obtained. Although it is known that many local judges continue to rule in favor of continuing life support in such cases, the risk management decision was to have the judge, rather than the patient, family, and physicians, make the ultimate choice. Experienced physicians and ethicists on the hospital's ethics committee unanimously concurred that the result of this administrative action will be that the hospital will always end up continuing to treat these patients, resulting in systematic overtreatment.

The fears that lead to the actions and current practices of risk managers appear to be largely unwarranted and, in many cases, entirely baseless. First of all, there is a total absence of legal liability resulting from withholding or withdrawing treatment from a patient. There has never been a successful criminal prosecution or a malpractice award against a physician or hospital in any case that fits this description. Either risk managers are ignorant of these facts or they are driven by other motives. A likely motive is the desire to avoid any and all litigation, not just the chance of losing a legal case.

Consider the calculations a risk manager would have to make in either situation. If calculations were based on the risk of losing a treatment withdrawal case that might be brought against the hospital, there would be less need for extreme conservatism because the probability of losing such a case is vanishingly small. Even if calculations were based on the risk of a criminal prosecution or malpractice suit being brought in the first place, the risk could be assessed as very low, even negligible. Few decisions made by anyone in other life situations are based on as low a probability of a negative outcome as those made by risk managers. This is why the very premise that underlies their decision making is irrational.

It is true that in the complex world of medical practice today, there is a need for orderly procedures and a decision-making process that is neither arbitrary nor potentially harmful to patients and is also just. However, it is implausible to hold that the best way to realize these ethical goals is to involve risk managers. It is even questionable whether these goals *can* be accomplished by individuals whose institutional mandate is geared not to the interests of patients, but rather to the business and public relations interests of the hospital. One noted authority disagrees, arguing that "in the vast majority of circumstances . . . risk management and health care ethics . . . ought to be compatible."[5] Despite this hoped-for congruence between the risk manager's role and adherence to ethical principles, the writer acknowledges "that those two roles . . . too often in practice turn out not to be congruous with each other."[6] If an idea is no good in practice, then it is no good in theory either.

The following cases offer a glimpse into the behavior of risk managers and lawyers in hospitals today. The first two cases were brought to an ethics committee for retrospective review after the episodes had occurred.

Rigid Insistence on Following a Rule

An obstetrician who had a long relationship with his patient was caring for her during a complicated pregnancy. The woman had one young child at home. At 34 weeks she experienced heavy active bleeding. A cesarean section appeared to be indicated, and because of the patient's underlying medical condition the physician advised her against having any more pregnancies. The patient accepted her physician's advice and agreed to undergo a tubal ligation at the time of delivery. She signed a consent form for that purpose when her serious medical condition became apparent. Thirty-six hours later, an emergency cesarean section had to be performed.

New York City has a regulation requiring a 30-day waiting period between the time a person signs a consent form for sterilization and the time the procedure may be performed. Some people contend that this regulation is paternalistic in its presumption that an adult who has been properly counseled requires the protection of a waiting period to prevent an impulsive decision and action. Regardless of the soundness of that criticism and the merits of a law designed to prevent abuses and erect safeguards, once a government regulation is in place, the hospital administration sets up procedures that must be followed. An exception to the 30-day rule is that of women who deliver prior to their anticipated delivery date. In that case, the regulation allows for sterilization to be performed sooner than 30 days but not sooner than 72 hours after the giving of the initial informed consent.

Aware of these regulations, the obstetrician consulted with risk management about the possibility of doing the tubal ligation despite the fact that only 36 hours had elapsed since the woman had signed the consent form for sterilization. The hospital administration would not allow the tubal ligation to be performed. The physician argued that the result would place the patient at increased risk, since he would then be forced to do a second surgical procedure at a future date, with the added risk to the patient of undergoing anesthesia a second time and the possibility that she could become pregnant in the interim, a life-threatening prospect. The hospital's department of legal affairs said: don't do the tubal ligation.

At the ethics committee meeting, a question was raised about the penalty for violating the city's regulation. The risk manager explained that all sterilizations must be recorded, including documentation of procedures following the 30-day or 72-hour rule. What would happen if a hospital violated one of these rules? The penalty would be a fine. Would it be justifiable to violate that rule, being prepared in advance to pay the fine and choosing to do so because that action would be in the best interest of the patient? This alternative was not in the repertoire of risk management.

I asked a different question about possible legal recourse: had anyone thought of calling a judge to obtain a ruling that would grant permission to do the sterilization under these circumstances? In general, it is not a good idea to go to court to obtain a prospective ruling every time a case raises some uncertainty about law or ethics. However, in this case and others like it, calling a judge opens the possibility of acting in the interest of the patient without violating the letter of the law. If a law or regulation that is generally designed to protect patients has the opposite effect in a particular case, asking a judge to suspend the rule for the particular case could be just the right course of action.

Finally, the risk manager was asked whether she knew of situations in which a hospital had violated one of these administrative regulations and petitioned the city agency for a waiver of the fine, based on the unique or extenuating circumstances of a case. She admitted that she did know of such cases, and that there were many different precedents established for departures from the rigid time constraints. The risk manager had not sought to avail herself of this mechanism, either through lack of imagination or an unwillingness to get involved in the extra tasks such petitioning would entail. The obstetrician, backed up by his other colleagues in the department of obstetrics and gynecology, was frustrated and angry at having been thwarted from acting as an advocate of his patient. The hospital's risk manager and the department of legal affairs had become enemies to be confronted, although it was clear that they might have found a way to get around the seemingly inflexible provisions of the law.

Denying an Abortion for a Comatose Woman

The patient was in a coma following an automobile accident and was three months pregnant. She was judged to have no reasonable likelihood of recovering her mental function, although she might live for years in the coma. The father of the fetus was out of the picture. The patient's mother, who had already granted informed consent for surgery on her incapacitated daughter, now asked that an abortion be performed. The hospital administration refused to allow the abortion, despite the fact that it was well within the legally permissible time limit and a physician was ready and willing to perform it. These administrators had seen no problem in allowing the comatose patient's mother to grant consent for risky neurosurgery, yet they were unwilling to honor her request to abort a previable fetus in her daughter, who would never regain consciousness.

The hospital attorney refused to seek a court order on behalf of the hospital and insisted that the patient's mother be the one to initiate a court proceeding, if that was what she wished. As these deliberations and delays occurred, the pregnancy drew dangerously close to the time of fetal viability, a point at which the legal and ethical concerns would loom as real barriers to procuring an abortion.

In whose interest was that action being thwarted? Although there are individuals in our society who proclaim that the fetus has a right to life from the moment of conception, no one connected with this case was making that moral claim. A hospital administrator had assumed decision-making authority that overrode both an incapacitated patient's family and hospital physicians willing to comply with the family's request. It is a generally accepted principle that family members may make medical decisions on behalf of their incapacitated relative, so long as those decisions are not clearly against the best interest of the patient. It is not plausible to argue that performing an abortion on a permanently comatose woman is against her best interest. Yet the issue of abortion is so highly charged in American society that it removes cases like this from the realm of medical ethics and thrusts them into the political arena.

The only way a refusal to grant the patient's mother's request for the abortion might be justified on ethical grounds is to insist that this case involves another "patient." Whether a previable fetus can properly be considered a patient, with rights and interests, is at least questionable and certainly controversial. It remains true that if the comatose woman miraculously were to wake up and request the abortion, she would have a legal and moral right to obtain it. In the absence of such a miracle, we are left to ponder the decision-making authority of the patient's next of kin—in this case, her mother. It is worth recalling that the hospital administration did not question the authority of the patient's mother to consent to neurosurgery on her daughter, a procedure that posed far more risk than an abortion. No one was requesting removal of the patient's life support. The request by the patient's surrogate (her mother, in this case, since there was no husband) was for abortion of a previable fetus. Fear of publicity and intimidation by the prevailing political climate led the risk manager to refuse to act. Although the motives of those who seek to act in the interest of the fetus are typically different from the reasons that propel risk managers, the trouble that foes of abortion might cause has succeeded in overwhelming all other concerns.

The final two cases in this series of illustrations received nationwide publicity. Both are significant for the reluctance of physicians to act in ways they believed to be ethically permissible, perhaps even ethically obligatory, because of the intrusion of hospital administration.

The Linares Case

A case involving a young child occurred in a Chicago hospital early in 1989. The patient was a 16-month-old boy who was irreversibly unconscious, ending up in a persistent vegetative state (PVS) for about eight months. The child, Samuel Linares, had aspirated a deflated balloon and was unconscious and blue when his father discovered him. The father, Rudolfo Linares, attempted mouth-to-mouth resuscitation, which was unsuccessful. Rudolfo Linares then called the emergency medical service and, while waiting for them to arrive, ran with the child to a nearby fire station. Firemen were similarly unsuccessful in attempting mouth to mouth resuscitation, but found the balloon in the child's throat and removed it. The child was then rushed to the emergency room of a nearby hospital, where he was found not to be breathing and having no heartbeat or pulse. After advanced life support measures were initiated, a pulse and blood pressure were established after at least 20 minutes, during which Samuel had no vital signs. He was then transferred from the community hospital to Rush-Presbyterian-St. Luke's Medical Center, a tertiary care hospital.

Despite early predictions by the pediatric intensive care staff that the child would die very soon, he continued to live in an unstable condition. A conference was held with the parents, and Mr. Linares said he wanted Samuel removed from the ventilator that was sustaining his breathing. The medical staff stated their belief that they could not disconnect the ventilator without a court order unless the child was brain dead, a condition Samuel Linares did not meet. However, a DNR order was agreed upon at the conference. The physician in charge of the pediatric intensive care unit, Dr. Goldman, and a pediatric neurology consultant agreed that the child had no reasonable chance for recovery of neurologic function. Despite Dr. Goldman's acknowledgment that he had disconnected ventilators from children in vegetative states on previous occasions, he deferred to the recommendations of the hospital lawyer. Rush hospital's legal counselor, Max Brown, told the physicians that the hospital would be civilly and criminally liable if the ventilator were to be disconnected.

The lawyer also said that a court could be petitioned for a declaratory judgment in the case. But he refused to agree to the hospital's petitioning the court on the grounds that the hospital might then be perceived as having a conflict of interest. Since Medicaid reimbursement for the care of the child did not cover the hospital's charges for treatment, it might look as if the hospital wanted the child removed from the ventilator because it was costing the hospital too much money. By April 1989, hospital charges were more than $600,000 and the hospital had been reimbursed for only $100,000.

Samuel's parents had repeatedly requested removal of the life support, and the hospital repeatedly denied their requests. Following months of frustration and failure to have his requests honored, Mr. Linares came into the hospital on one occasion and disconnected the lines and tubes providing life support for his son. A nurse quickly reconnected them. Four months later, plans were being made to transfer Samuel to a nursing home, and a nurse left a recording on the Linares' answering machine saying that the transfer was to occur the next day. Mr. and Mrs. Linares arrived at the hospital shortly thereafter to visit Samuel in the pediatric intensive care unit. Rudolfo sent his wife back to the car and drew out a pistol. He then pulled the plug of the respirator that had been keeping his son alive, saying, "I'm not here to hurt anyone." He held the child in his arms until Samuel died.

A state's attorney declared that "the facts of this case clearly dictate the filing of first degree murder charges." The coroner said that the balloon was the "primary" cause of death but drew no conclusion about whether the "manner of death" was an accident or a homicide. A grand jury later refused to indict Rudolfo Linares for murder. He received a suspended sentence on a weapons charge, which was a misdemeanor. There was heated public discussion surrounding this case, and much sympathy was expressed for the Linares family.[7] One commentator on the Linares case noted that "Mr. Linares' final outraged response was, if not praiseworthy, at least understandable."[8] Another observed that the father's desperate act is one "we do not urge others to emulate, but that we understand. It was an act we should not say was right, but we cannot in our hearts condemn as ultimately wrong."[9]

How can an act be "not right" and yet "not ultimately wrong"? This conclusion appears almost contradictory, yet a plausible interpretation can be given. The argument is not that patients or their families should be encouraged or even permitted on a routine basis to settle disputes with physicians or hospitals by threatening with weapons or using physical force. Yet when a family acts out of rage or frustration in a situation resulting from defective legal advice, it is not inconsistent to judge their action to be morally excusable under the extenuating circumstances, though in general wrong.

It is important to underscore the fact that the Linares case did not involve a conflict of ethical beliefs on the part of the child's parents, the pediatrician in charge, or even the hospital attorney. Dr. Goldman, the attending physician, said, "There was no ethical difference of opinion here. The physicians agreed that the child was in an irreversible coma and would not recover. There was no medical opposition to removing the ventilator. What we faced was a legal obstacle."[10]

Scholars familiar with the law regarding withholding and withdrawing life support contend that Max Brown was simply wrong in his judgment of what the law could allow. One article points out that statutes in the state of Illinois support a decision to terminate treatment. In addition, "Though Illinois case law does not address this issue, precedents from other states suggest that state courts will support decisions to terminate treatment of children in a persistent vegetative state."[11] Perhaps more important: "*No physician has ever been convicted of any crime for withholding or withdrawing treatment from such a patient, in spite of much evidence in the medical literature that such decisions take place.*"[12] It is therefore truly remarkable that hospital lawyers and risk managers regularly assert that there is a danger of criminal liability in such cases. To apply the epithet "paid paranoids" to hospital attorneys and risk managers does seem fitting.

Another commentator described Max Brown's actions as "distorted and detective legal advice that led . . . physicians and Rush to a not uncommon form of moral paralysis. . . ."[13] This serves as a reminder of the admonition that physicians should know at least enough law to be able to tell when the advice their lawyers are giving them is so incredible that it is most likely wrong. Brown argued that withdrawing the ventilator from Samuel Linares could be considered first-degree murder. He instructed Dr. Goldman to inform Mr. and Mrs. Linares that removing Samuel from the ventilator "was not a legal option." This information could only confuse parents who were also told that the hospital would not seek a court order, but they could do so themselves if they wished.[14] How could this couple understand a statement that to stop the ventilator would be murder, when they were also being told that they could ask a judge to grant permission to do the murderous deed?

The obstacles created by Rush Hospital to removing a child in PVS from a ventilator were based on unsound legal advice. Moreover, as critics uniformly observed, this sort of action on the part of the hospital is unethical "because it places the self-interest of the physician (or hospital) above that of the patient, and it ignores the fact that responsibility must be assumed by professionals who operate in a radically imperfect world."[15] Now it might be replied that Samuel Linares, like all patients in coma or PVS, no longer had any interests. He could neither be benefitted nor harmed by being kept on a ventilator or other life support indefinitely. The same cannot be said for his parents, however. They continued to suffer as long as the child they loved and once saw as a vital, functioning human being was being kept biologically alive by machinery.

Moreover, not only did the hospital's actions cause emotional harm to Mr. and Mrs. Linares, their rights as decision makers for their child were violated. Law and ethics both grant decision-making authority to parents, so long as their decisions do not result in "medical neglect" from a legal point of view and are not clearly against the best interest of the child from an ethical standpoint. Neither state statutes, case law, nor the federal Child Abuse Amendments of 1984 prohibit removal of a ventilator from an irreversibly comatose child. [16] And it could not plausibly be argued that it is against an irreversibly comatose person's interest to remove machinery that is keeping the heart and lungs operating.

Thus, with no ethical opposition on the part of anyone to removing a ventilator from Samuel Linares, the hospital's attorney refused to allow withdrawal based on a *theoretical* possibility of exposure to legal liability. As others have correctly pointed out, attorneys should give advice based on probabilities, not abstract possibilities. "The practical probability in this case of a physician being prosecuted for murder was remote to the point of being nonexistent, fundamentally because the law and the facts of the situation simply would not support even a vaguely plausible prosecution for unlawful, malicious killing."[17]

The mentality of risk management is reflected in the quest for 100 percent immunity. Although physicians cannot be expected to be fully conversant with laws, they can be expected to use common sense. In cases like the Linares case, common sense means questioning judgments such as those issued by hospital lawyer Brown. It's not apparent why physicians generally act as though they do not have the option of seeking a second legal opinion when the case they are managing occurs entirely within a hospital. One reason could be that the lawyer for their hospital refuses to immunize them against potential liability, and doctors would then refrain from acting without the protection afforded by the hospital's legal counsel.

The physicians in charge of Samuel Linares perceived no ethical dilemma in removing the ventilator. Nor did they see it as an ethically ambiguous situation. Can ethical judgments be so far removed from legal mandates that they come into direct conflict? That is unlikely, and it is something physicians should reflect on when they run to risk management for guidance and blindly follow the advice of the hospital attorney.

As for what happens to physicians who fail to comply with directives from the hospital administration, informed opinions vary. Some physicians imagine a worst-case scenario, such as summary dismissal from the hospital staff. Others contend that immediate dismissal is not likely to occur, but that physicians who ignore or fail to consult risk management will be put on notice and their actions carefully monitored. The prospect of bad consequences seems greatest for junior physicians, and it is worthwhile to speculate on what would happen if powerful figures such as department chairmen formed an alliance to resist intrusions by administrators.

Nathan Hershey, a professor of health law writing about the Linares case, comes to the defense of Max Brown. Hershey faults Brown's critics for failing to understand the pressures on a hospital attorney and for ignoring the fact of who the attorney's client is. The attorney's client is the hospital. "Patients, and the physicians on the hospital staff who are not employees, are not clients of the attorney. . . . [T]he content of the legal guidance is controlled by the attorney–client relationship."[18] These observations are proper reminders of the obligation lawyers have to their clients. But that point cannot serve to justify a system whereby the interests of hospitals are placed above those of patients and families.

Hershey explains the actions of Rush Hospital and its attorney, Max Brown, and in so doing defends what is clearly an indefensible system of preventive legal maneuvers and public relations ploys: "to assume that casting the potential for legal involvement in terms of probabilities, rather than possibilities, would ordinarily satisfy physicians, is to be unrealistic in many situations."[19] Is it better, therefore, to be realistic with regard to satisfying physicians' worst nightmares yet remain *unrealistic* with regard to the actual likelihood of adverse legal outcomes? Instead of giving in to the demand of those physicians who seek absolute guarantees, lawyers and risk managers would do better to help shape a climate in which practicing sound medicine and establishing good physician–patient relationships are the preeminent goals.

In opposition to Hershey's view, another attorney characterizes the risk management counsel given to Rush hospital as "ultraconservative."[20] This critic argues that "risk management advice based mainly on imagined, hypothetical legal demons rather than a realistic, pragmatic appraisal of actual legal risks directed the hospital to behave in a manner that contradicted the ethical interests of the patient and family, the hospital itself and its medical and nursing staff, and society."[21] In other words, ultraconservative risk management, at least in the Linares case, resulted in no one's interest being served, not even the hospital's.

As for the public relations aspect, it is dismaying to find a professor of health law justifying a hospital's actions because of the potential for adverse publicity. Hershey writes:

> Termination of care for a patient without private insurance or personal resources may raise concern in hospital management about misinterpretation of motives by the media. A hospital that facilitates a patient's family obtaining a court order to terminate care may become suspect. Perhaps, ideally, media attention should not be a factor in health care decision-making, but it sometimes is.[22]

The prediction about "media attention" backfired in the Linares case. How could the hospital have foreseen that a distraught father would take matters into his own hands, entering the hospital with a gun and removing his comatose infant from the respirator? Although people were no doubt troubled at the prospect of gun-toting parents invading pediatric intensive care units, they could readily sympathize with parents whose wish to withdraw life support from their irreversibly unconscious child was thwarted by misguided bureaucrats at the bedside.

The Case of Angela Carder

The second case that received national publicity occurred in a hospital in Washington, D.C., in 1987. Angela Carder was a 27-year-old married woman who, in her dying days, was forced to undergo a cesarean section to try to salvage her marginally viable fetus.[23] She had recently been diagnosed as having a recurrence of cancer from which she had suffered intermittently since the age of 13. She had intentionally become pregnant in her desire to start a family and had completed about 26 weeks.

The person who had oversight of the medical center's risk management program at the time was a woman with a master's degree in health care administration.[24] On the morning of June 16, 1987, she was summoned to the office of the director of administrative affairs of GW Hospital, to whom she reported. The director told the risk manager that "we had a patient, who was close to death with cancer and was pregnant. And the OB/GYN staff was questioning whether to go along with the family's idea to let the baby die, once this patient dies."[25] The risk manager later told a lawyer:

> [The director] made some comments about how the courts had historically maintained what he recalled from all of his training and experience was that 26 weeks was in the definition of "viable" from a legal perspective. He felt that complicated the situation and felt that we should get advice as to whether we had a legal situation on our hands and asked me to contact general counsel. We also, at that point, did indicate to the physicians that we felt that until we could get in touch with general counsel . . . they should proceed to get ready for a postmortem section, given that her death was presumed to be imminent. We really didn't know how much time we had to work with and they were indicating to us an hour to two hours.[26]

When the risk manager conveyed to the hospital's general counsel the facts surrounding Angela Carder, one of the attorneys said they should call a judge to figure out what course the hospital should take.[27] The lawyer from the general counsel's office then got in touch with a judge, who decided to hold the hearing.

The Angela Carder case raises two different but equally serious ethical concerns. The first is the procedural question: Who should decide whether to undertake invasive medical procedures when the patient is pregnant? The second, more controversial question is one of ethical substance: Does a fetus in utero have moral standing that can override the rights of the pregnant woman to decide about her own medical treatment? At George Washington University Hospital, administrators answered the first question by disqualifying the patient, her family, and her physicians from any decision-making role. The proper decision maker could only be a judge. As it would later turn out, this procedural step was determined to be legally flawed as well as ethically mistaken. To attempt to answer the substantive ethical question, we need to plunge headlong into the issue of maternal–fetal conflict.

Notes

1. Marshall B. Kapp, "Are Risk Management and Health Care Ethics Compatible?" *Perspectives in Healthcare Risk Management*, vol. 11 (Winter 1991), p. 2.

2. George J. Annas, quoted in Lawrence J. Nelson and Ronald E. Cranford, "Legal Advice, Moral Paralysis, and the Death of Samuel Linares," *Law, Medicine & Health Care*, vol. 17 (Winter 1989), p. 322. Nelson and Cranford state that they "agree completely" with Annas's assessment of this phenomenon.

3. Kapp, "Risk Management," p. 2.

4. Ibid.

5. Ibid., p. 6.

6. Ibid., p. 2.

7. The above description and facts about the Linares case are taken from John D. Lantos, Steven H. Miles, and Christine K. Cassel, "The Linares Affair," *Law, Medicine & Health Care*, vol. 17 (Winter 1989), pp. 308–15.

8. Ibid., p. 314.

9. Nelson and Cranford, "Legal Advice," p. 317.

10. Cited in ibid., p. 316.

11. Ibid., p. 314.

12. Ibid; emphasis added.

13. Ibid., p. 316.

14. Ibid.

15. Ibid., p. 322.

16. Ibid., p. 320.

17. Ibid., p. 321.

18. Nathan Hershey, Letter to the Editor, *Law, Medicine & Health Care*, vol. 18 (Winter 1990), p. 425.

19. Ibid.

20. Kapp, "Risk Management," p. 3.

21. Ibid.

22. Ibid.

23. *In re A.C.*, 533 A.2d 611 (D.C. 1987). The account that follows is taken from the trial court transcript in the case; from an appeals brief filed by the court-appointed attorney for the fetus; from an *amicus curiae* brief of the American Medical Association, the American College of Obstetricians and Gynecologists, and the Medical Society of the District of Columbia; and from the Opinion of the Court in the Hearing En Banc of the

District of Columbia Court of Appeals (argued September 22, 1988, and decided April 26, 1990).

24. The material that follows is taken from the deposition taken by lawyers in the family's suit that followed the original judicial hearing.

25. Deposition of St. Andre, p. 24.

26. Ibid., pp. 32–33.

27. Ibid., p. 89.

4

The Fetal Police: Enemies of Pregnant Women

When court orders are sought by physicians or hospitals on behalf of fetal "patients," they have usually been granted by sympathetic judges. Many judges have adopted the procedure of balancing fetal and maternal interests. When the state engages in such balancing, it treats a woman and her fetus as co-equal combatants before the law.

The case of Angela Carder illustrates an ethical problem related to procedures: how a hospital administration and its lawyers can insert themselves into the doctor–patient relationship. The case involved another ethically questionable action, one that violated substantive as well as procedural ethics: performing a court-ordered cesarean section on a nonconsenting woman with a marginally viable fetus.

The situation was fraught with a number of uncertainties: uncertainty about the actual wishes of Angela regarding consent to a cesarean section; uncertainty about her capacity to grant genuinely informed consent in her condition of terminal illness and under heavy sedation; and uncertainty about the chances of survival or serious impairment of this marginally viable fetus. However, no uncertainty should exist regarding the need to obtain a patient's informed consent before performing surgery on her. Nor should there be any question of the right of a pregnant woman—even one who is terminally ill—to refuse a cesarean section.

At the time these events occurred, Ms. Carder's health was rapidly deteriorating. X-rays revealed an inoperable tumor in her lung. When she entered the hospital, she was asked if she really wanted to have the baby, and she replied that she did. She agreed to treatments intended to extend her life until at least her 28th week of pregnancy, at which time the potential outcome for the fetus would be greatly enhanced. As her condition worsened, her physicians estimated that she was unlikely to live for more than 24 hours. When asked again if she still wanted to have the baby, Ms. Carder was somewhat equivocal in her reply. The fetus was then in "a chronically asphyxiated state," which increased the chance that the fetus would die or, if it lived, would suffer permanent handicaps.

It was at this point that the hospital decided to go to court for a ruling on what was legally required. The trial court in the District of Columbia appointed a counsel for the fetus, and the District of Columbia was permitted to intervene on behalf of the fetus in the state's role as *parens patriae*—a role designed to protect vulnerable individuals unable to speak for themselves. On June 16, 1987, a hearing was hastily convened in the hospital, with different attorneys representing the hospital, the fetus, and the District of Columbia; a lawyer for Angela Carder, the patient; several physicians from the hospital; Angela's mother, who testified; and her husband, who was too distraught to testify.

When patients are unable to grant consent for themselves, an appropriate surrogate (usually a family member) is asked to consent to surgery recommended by physicians as being in the best interest of the patient. In the case of Angela Carder, no family member stepped forward to consent to a cesarean section. Her husband stood by her, her parents urged that the cesarean should not be done, and Angela's own physician was willing to abide by her apparent wishes, elicited while the hearing was still going on, not to have the surgery. During the course of the hearing, one of Angela Carder's physicians stated that the doctors did not want to perform a cesarean section on this patient.

Given this array of uncertainties, it might be supposed that a court order to proceed with a cesarean section could be justified. But a close scrutiny of the facts and circumstances shows that supposition to be ethically wrong and legally questionable. The physicians caring for Angela Carder testified at the hearing that a cesarean section should not be performed because they did not believe that either Ms. Carder or her family wanted it. However, the judge concluded that the patient had not clearly expressed her views in the matter, and that because she was now unconscious, she was unable to do so. So the judge ordered the surgery to be performed. That judicial maneuver was at least peculiar and at worst violative of the patient's rights. Lack of clarity or inability to grant consent should result in an invasive procedure's *not* being performed, rather than creating a presumption in favor of performing it.

In response to a question posed by a lawyer who took depositions in the case, the hospital's risk manager acknowledged that she thought that letting the baby die would not be a good thing. She admitted that it was probably her own "emotional and personal face rather than my institutional face showing."[1] In her deposition, the risk manager stated:

I have thoughts about possible hospital liability in every situation because that is what I am paid to do. In this particular situation, I don't think liability was the issue that was governing our process. . . . In this case there was enough uncertainty that general counsel needed to be consulted. . . . I think that we just needed to make sure that we weren't dealing with an issue that had legal overtones to it. And that was the reason for consulting counsel, just to get that clarification.

The main thing was the fact that there was a fetus that was presented to us to be viable and that we could potentially be facing a situation where a living infant was born and someone was telling us to let it die.

The fact that there may be legal overtones is to me the trigger that counsel needs to be consulted.[2]

The risk manager's boss, the director of administrative affairs, was also deposed in the Angela Carder case. The following exchange between the director and the deposing attorney demonstrates his view, as a hospital administrator, of what it was necessary and proper to do:

"I said my basic interpretation of the situation is that we have a viable fetus and this was a societal question, not one for us to answer and that the institution was in a position where it should seek outside advice on this."

"What do you mean by societal question?"

"It is a question of the rights of the newborn baby, of the unborn baby."

"Would that be a question that would exist if the parent or the mother was competent to make a treatment decision?"

"Yes."

"Did you view that as raising a legal question?"

"Yes."

"Did you view it as raising a moral question?"

"No."

"Not personal values?"

"No personal value. . . . In graduate school we go through a certain amount of legal training . . . and it's just broad census in a lot of the case law. In *Roe v. Wade* it was our basic guideline that at 24 to 25 weeks it is a little gray but beyond 25 weeks it is a no-abortion sort of situation.

"I realized, one, that I had been told by a clinician that we had a probable viable fetus and that she was 26 weeks of gestation.

"I'm not going to make judgments for the institution when I think society is wrestling with those very same issues. That's why I saw it as a question for general counsel as to what do we do.

"He saw it obviously it was a question for society through the courts."

"Are you aware of other situations in the hospital where a pregnant woman experienced compromising conditions that would adversely affect fetal well-being but that intervention was not indicated for the health of the mother, whether that would raise the same kind of concerns that would require the involvement of counsel?"

"If it was raised to me, yes, it would require me to seek advice of general counsel."

"Did the hospital believe that it might face liability if it failed to perform a Caesarian section for the fetus if Angela Carder had refused the surgery at that stage of her pregnancy?"

"I'm sure that occurred. I don't recall that as a major consideration at the time. . . . liability . . . wasn't the major concern.

"Liability in the health industry is always in the back of your head. But liability
was not discussed to my knowledge."[3]

The medical director of George Washington University Hospital was also
deposed in the case. When asked by the lawyer why it was necessary to have
the court involved when the family members were available to make the
decision concerning a cesarean section, the medical director replied that
"viability was a serious issue here." He added that it was not within the
wisdom or province of either the family or the hospital to make that decision.
"It's in the courts," he stated, "It's a societal decision."[4]

Curiously, however, the medical director did not refer to cases of cesarean
sections that were brought to court for a judge's ruling, but rather to *Roe v.
Wade*. Neither the medical director nor the director of administrative affairs
appears to have noticed that *Roe v. Wade* was a U.S. Supreme Court decision
guaranteeing a woman a constitutional right to abortion up to the point of
fetal viability, while Angela Carder's situation involved her or her family's
right to refuse unwanted surgery. It was quite clear, however, that the medical director did not believe it to be the right of patients or families to make
any decisions where a viable fetus was involved: "Who decides about whether
a fetus is given an opportunity to live? Is that a family decision or, again, one
that society must decide for us through the courts."[5]

As a final affront to Angela's human dignity, the argument was offered that
this patient was dying anyway, so it didn't really matter if her life would be
shortened in an attempt to salvage her fetus. The court-appointed attorney
for the fetus began her argument by stating that "we are confronted . . . with
a need to balance the interest of a probably viable fetus, a presumptively
viable fetus, age 26 weeks, with whatever life is left for the fetus's mother."
The attorney questioned "what we will be depriving her [Angela] of realistically if we were to take measures to protect the life of the fetus at this
point."

Later in the hearing, the lawyer representing Angela Carder stated an
objection to this line of reasoning: "As I understand the medical testimony, if
we were to do a C-section on this woman in a very weakened medical state,
we would in effect be terminating her life." The judge's reply was: "She's
going to die."

Moments later, however, the judge asked the lawyer appointed for the
fetus: "Why should I possibly hasten the mother's death by ordering the C-
section?" The lawyer answered, "well, I suppose it will hasten her death but
her mother testified that she asked to be relieved of the burden last evening,
that she had enough of the pain. She only wanted to be relieved of it."
Perhaps this was an argument in favor of increasing Angela's pain medication
or of sedating her to the point where she would lapse into unconsciousness. It

could even have been the beginning of an argument supporting active euthanasia. It does not, however, stand as a plausible defense of a decision to perform surgery against the wishes—albeit expressed unclearly—of a patient in extremis.

The judge at the hearing asked the same question of another attorney, one who represented the city (Washington, D.C.) in arguing for "a compelling state interest" in preserving the life of a fetus. The judge asked: "why should I possibly hasten the death of the mother by ordering a C-section in an effort to allow the fetus to possibly live?" The lawyer replied that "it's a balancing situation, as in many situations. . . . I think balancing her ultimate end, death within a very short period of time, the risks to the fetus by delaying . . . I think we have no choice under that circumstance but to perform the cesarean section. Balancing those factors, I think the potentiality of this fetus outweighs the imminent death of the patient."

I contend that all of these attorneys—the lawyer representing the city, the hospital's lawyer, and the court-appointed lawyer for the fetus—were mistaken in applying a balancing test to this situation. Any attempt to balance the right of a woman to refuse surgery, or her right not to have surgery performed without her informed consent, against the alleged interests of a marginally viable fetus is morally obtuse. Balancing the interests of the fetus against the right of a pregnant woman to refuse an invasive procedure is ethically inappropriate, if not a legal fiction designed to come out in all cases in favor of a fetus. This is because a balancing test will inevitably weigh the right to life more heavily than the right to informed consent.

After the order was issued but before the conclusion of the hearing, one of the doctors went to Angela Carder's bedside and found that her sedative had worn off. The physician was able to communicate with her, and when she was told of the court's order, she said she would agree to the cesarean section. However, about 20 minutes later, the physicians returned to the bedside with Angela's family in order to verify her consent to surgery. The physician testifying told the judge that Angela was at that time "responding, understanding, and capable of making such decisions."[6] He testified that Angela asked her own physician whether she would survive the operation. She asked him if he would perform it. Her doctor told Angela he would perform it only if she authorized it, but it would be done in any case. The physician at the hearing then testified that Angela "very clearly mouthed words several times, I don't want it done. I don't want it done." The judge concluded that he did not know what Angela Carder wanted and reaffirmed his order to perform the cesarean section.

The surgery was performed, and a baby girl was delivered. The infant died within two and one-half hours, and Angela Carder died two days later. The hospital's attorneys and risk manager who sought the court order, the attorney who represented the District of Columbia, the lawyer for the fetus, and the judge who ordered the cesarean section all acted as enemies of this dying patient. In subsequent proceedings, the original court order was vacated, meaning that it would not stand as a judicial precedent, although, of course, the circumstances could not be reversed. The District of Columbia Court of Appeals stated:

> If the patient is incompetent or otherwise unable to give an informed consent to a proposed course of medical treatment, then her decision must be ascertained through the procedure known as substituted judgement. Because the trial court did not follow that procedure, we vacate its order. . . .[7]

This conclusion of procedural justice was buttressed by a further observation made by the court of appeals. The court asserted "that it would be far better if judges were not called to patients's bedsides and required to made quick decisions on issues of life and death."[8] That is precisely what occurs in most situations where hospitals seek court orders to override treatment refusals by pregnant women.

The court of appeals rejected the notion that the interests of the fetus must be balanced against the right of a pregnant woman to refuse an invasive procedure. The appeals court also dismissed the argument that under the substituted judgment procedure, Angela Carder would have consented to the cesarean section. The court declined, in addition, to accept the contrary arguments that Angela was competent and made an informed choice not to have the cesarean performed, or that even if the substituted judgment procedure had been followed, the evidence would necessarily show that Angela would not have wanted the operation. The appeals court opined: "We do not accept any of these arguments because the evidence, realistically viewed, does not support them."[9]

The opinion by the court of appeals reiterated a number of well-established principles in biomedical ethics and law. The "analysis of this case begins with the tenet common to all medical treatment cases: that any person has the right to make an informed choice, if competent to do so, to accept or forego medical treatment. The doctrine of informed consent, based on this principle and rooted in the concept of bodily integrity, is ingrained in our common law."[10]

A second significant principle articulated by the court of appeals is that "courts do not compel one person to permit a significant intrusion upon his or her bodily integrity for the benefit of another person's health."[11] The court's opinion cited a case requiring consent by the parents of a 15-year-old for a skin graft that would benefit a severely burned cousin.[12] It also cited the case of *McFall v. Shimp,* in which the court refused to order Shimp to donate bone marrow necessary to save the life of his cousin, McFall.[13] Considering the argument that fetal cases are different from these others because they involve enhanced duties on the part of a woman who "has chosen to lend her body to bring a child into the world," the court rejected that argument. "Surely . . . a fetus cannot have rights in this respect superior to those of a person who has already been born."[14]

Maternal–Fetal Conflict

Pregnant women are at risk for loss of their rights as patients from the actions of physicians (who should be their advocates), risk managers (who shouldn't be involved at all), and judges (who may become involved whether they want to or not). They may also suffer loss of their rights by a statute or regulation, which can have the effect of limiting the rights of patients or creating conflicts with rights that have already been established in other sectors of the law. Not all laws are put in place with the aim of safeguarding the rights of patients. Recall the "gag rule," a U.S. government regulation that prohibits personnel in health care facilities that receive federal funds from discussing with a pregnant woman her right to an abortion.

It is crucial, however, to distinguish the ethical arguments relating to maternal–fetal conflict from those that surround the abortion debate. The issues in each controversy are separate and distinct. The two situations were conflated and confused in the original trial court case of Angela Carder. The abortion controversy centers on the moral status of the conceptus, embryo, or fetus. To resolve this controversy, opponents have to reach agreement on the point following fertilization at which the product of conception acquires rights—in particular, the right to life. No such agreement need be reached in the arena of maternal–fetal conflicts. There is widespread agreement—in fact, almost universal acknowledgment—that once born, infants have a life that deserves protection from harm. The ethical question here focuses on what steps may be taken to seek to ensure that infants are "well-born," that is, born as sound and healthy as possible.

A further point is worth noting in distinguishing maternal–fetal conflicts from the abortion controversy. In the context of abortion, as the pregnancy comes closer to term, the ethical concerns grow correspondingly. In contrast, if the cause for concern is damage to the fetus likely to result in anomalies or developmental disabilities after birth, rather than the life of the fetus itself, the early weeks and months are often the most critical time. Scientific evidence regarding the kinds of serious, irreversible damage that can be inflicted very early in pregnancy suggests that some of the worst consequences for infants can result from things that take place possibly even before a woman realizes that she is pregnant. The priorities that exist in the abortion context are almost reversed in maternal–fetal conflicts where the pregnant woman is abusing drugs or alcohol.

Sometimes it is a doctor—usually an obstetrician—who makes the fetus an enemy of the pregnant woman. Obstetricians who seek court orders to override a woman's refusal of a cesarean section appear to place a higher value on the fetus than on respect for the rights of their autonomous adult patient. In addition, obstetricians may have other, self-interested motives. Their desire for the best possible outcome may well stem from their desire to be free of future liability and criticism. In no other medical situation is surgery done on an unwilling patient for the sake of another person. No court has ever ordered a parent to donate a kidney for transplantation in order to save the life of the child. Nor have physicians gone to court to seek such coercive intervention. Even a renewable bodily resource—such as bone marrow—is never ordered by courts from an unwilling person in order to benefit a relative. Yet in the hope of preventing possible harm to a fetus, doctors have requested and judges have granted permission to operate on women and to perform blood transfusions in violation of their religious prohibitions.[15]

One justification physicians use in seeking to control pregnant women is that the fetus, in addition to the woman, is a *patient*. This reference to two patients—the pregnant woman and the fetus—creates an aura of conflict. One physician writes that these cases of conflict "involve a struggle between valid obligations to two patients, obligations that very rarely are in opposition."[16] However rare such cases of conflict may be—and they appear to be increasing—when the interests of one patient conflict with the rights or interests of a second, the two become potential adversaries.

In general, adult patients have a legal and a moral right to refuse medical treatment, even if their reason for refusing is thought to be irrational. When a patient's mental status is questioned, an evaluation is typically done to ascertain whether the person lacks the capacity to make medical decisions. Although the concept of "competency" is in its strict meaning a legal notion, it is used informally in the medical context to denote a patient's capacity to grant informed consent to treatment. There is widespread agreement that patients are considered competent unless they have been determined to be incompetent, even if their choices appear irrational to a physician or to others.

Physicians have recommended and sought to impose a number of different medical or surgical interventions on pregnant women, sometimes for the woman's own benefit and almost always for the sake of the fetus. Probably the most common of these is cesarean section, which women have refused on religious grounds, because of fear of cutting, or for other reasons—some rational, others irrational. The practice of treating adult pregnant women differently from other competent adults stands in need of ethical justification.

Cesarean sections are not the only recommended intervention about which conflicts occur. A pregnant woman may be forced to take medication, such as penicillin, for the sake of fetal health. In a case that occurred in 1982, a court ordered a pregnant diabetic woman to receive insulin treatment despite her refusal on religious grounds.[17] For many years, Jehovah's Witnesses have been compelled to receive unwanted blood transfusions, including transfusions performed well before the onset of fetal viability. In a case that occurred in Jamaica Hospital in New York in 1985, a court ordered a Jehovah' Witness to be transfused when the fetus was only 18 weeks in gestation.[18]

In addition to recommending and obtaining legal coercion for specific medical interventions, physicians have sometimes sought more generally to compel a woman's compliance during pregnancy. Doctors' orders have included putting on weight or limiting weight gain, eating particular foods, taking vitamins and other medications, not carrying heavy groceries, making and keeping doctors' appointments,[19] and refraining from having sexual relations.[20]

Still the most common situations of forced treatment are those involving recommended cesarean sections. The following case has some unusual details, but is not uncommon as an illustration of the typical practice. The patient was a 28-year-old woman brought to the hospital emergency room by ambulance in active labor. The woman had gestational diabetes during pregnancy but was otherwise in good health. She had had four previous deliveries of large babies, ranging from 10 to 14 pounds. Now she appeared at the hospital and was assessed as needing a cesarean section. She refused. At the time, she stated that she was opposed to surgery because of her religious beliefs, but she later said that she and her husband "didn't believe in medicine." The doctors told the patient that the cesarean section was necessary for the well-being of the fetus. Also, the fetus was estimated as very large and might not come out. In that case, the woman herself would be at risk of death.

While the woman continued to refuse, doctors sought a court order for a cesarean section. The judge had been called and was on his way to the hospital when the patient finally consented—under duress—to the cesarean section. Everyone acknowledged that the pressure of the judge about to arrive was what led her to change her mind. A healthy, 14 ½-pound baby was born. Although that was the best possible outcome, the means to achieve it remain ethically questionable. To seek to persuade a woman to undergo a surgical procedure for the sake of the fetus is surely ethically acceptable; to coerce her is not.

When the case was discussed retrospectively at our conference, several young doctors agreed that they were left with no choice in the matter. One said, "You *must* save the fetus in this situation." Another confessed that if the

woman continued to refuse, he would give her an injection to put her to
sleep and then do the section. "You have to do it," he insisted. The consensus at the conference was that "the rights of the fetus trump the rights of the
mother."

The views expressed by the obstetrical residents at this conference stand in
sharp contrast to a statement issued by the Committee on Ethics of the
American College of Obstetrics and Gynecology (ACOG). That statement
said, in effect, that it is ethically unacceptable to coerce pregnant women and
force them to undergo cesarean sections or other recommended procedures.
The concluding paragraph of the statement reads as follows:

> Obstetricians should refrain from performing procedures that are unwanted by a
> pregnant woman. The use of judicial authority to implement treatment regimens in
> order to protect the fetus violates the pregnant woman's autonomy. Furthermore,
> inappropriate reliance on judicial authority may lead to undesirable societal conse-
> quences, such as the criminalization of noncompliance with medical recommenda-
> tions.[21]

However, actual practice and physicians' attitudes deviate considerably
from the urgings of the ACOG Ethics Committee's statement. An article
published only five months before the statement was issued reported on a
survey of heads of fellowship programs in maternal–fetal medicine and directors of residency programs in obstetrics and gynecology. This article revealed
that 46 percent of the heads of fellowship programs thought that women who
refused medical advice and endangered the life of the fetus should be detained. In addition, 47 percent supported court orders for procedures such as
intrauterine transfusions. It is clear from these statements and articles in the
professional literature, as well as from informal discussions among physicians, lawyers, and ethicists, that physicians are divided and conflicted about
their moral and legal obligations in these matters.

The case of the woman with gestational diabetes described above would
not be reported officially, since the court order proved unnecessary. Yet the
distinction between this case and court-ordered cesareans is a difference only
in legal reporting. About 20 court-ordered cesareans have been reported in at
least 14 different states and the District of Columbia.[22] It is estimated that a
much larger number have gone unreported and that such cases occur with
some regularity. In another instance, the hospital administration went to
court prospectively to obtain a judicial order, which was granted. The patient
then left the hospital and hid from the doctors. She subsequently delivered
by normal vaginal delivery, and the baby was fine.

Still another case involved a 31-year-old patient who came in at 26 weeks
for assessment because of decreased fetal movement. Mrs. G had had four

pregnancies: three spontaneous miscarriages and one live birth, a 2½-year-old at home. Diagnostic studies of umbilical blood flow revealed an ominous sign for fetal well-being. The risks to the fetus were explained to Mrs. G, who was told she needed prolonged bed rest, continued hospitalization, and close fetal surveillance. The estimated weight of the fetus was 633 grams, which suggested intrauterine growth retardation. Mrs. G became very upset, saying she had to go home to her other child.

Some of the doctors at the conference said that she was "unreasonable," that she was denying, and one even asserted that she was "hysterical." Others observed that she was genuinely and properly concerned about her child at home. Having been told that her 26-week fetus had a bleak prognosis, she was more concerned about caring for her healthy child. Mrs. G was reported to have said, "They told me my baby was going to die, so why should I stay here?" One resident noted that this was what the patient heard, but that it wasn't stated in precisely those terms.

Some participants stated that the woman was hysterical and therefore lacked decisional capacity; others argued that she was highly emotional yet still capable of making a decision. Discussion also focused on whether the infant's poor prognosis, even if the patient followed medical advice, was a reason to respect her refusal of the doctor's recommendations; whether there were any legal grounds for seeking to keep her in the hospital against her will; whether she should be permitted to refuse a cesarcan section if that was deemed necessary in the event of fetal distress; and whether a damaged baby might be a worse outcome than fetal death. The conference ended with the information that Mrs. G agreed to amniocentesis and to steroid administration to aid fetal lung maturation, but she still insisted on leaving the hospital against medical advice.

Reasonable people disagree sharply about what is the ethically right thing to do in cases such as these. Some question whether it matters what reasons the patient gives for refusing a treatment recommended for the sake of the fetus. Does gestational age matter, and if so, what should the cutoff point be? Other critical questions focus on the prognosis: Is it less justifiable to force women to have a cesarean section or other interventions when the prognosis is poor than when the fetus appears to be normal and healthy?

At one extreme are those who argue in favor of intervention because it is the obligation of the obstetrician to protect the "vulnerable" patient, the fetus, which is incapable of decision making. At the other extreme are those who argue that there is no acceptable ethical justification for forcing mentally competent women to receive medical or surgical treatment, or for detaining them in the hospital, even when they are pregnant. The fetus is contained entirely within the woman's body, so to intervene on behalf of the fetus requires the use of legal coercion, physical force, or chemical restraints.

Courts have long been involved in decisions surrounding medical care and treatment. In many instances, judges have been called upon to act in the traditional role of courts as adjudicators of disputes. When a physician seeks to overrule a patient's refusal of treatment, an appeal is sometimes made to the judicial system. In these refusal-of-treatment cases—or, in the phrase used to describe a subset of cases, the "right to die"—judges have set important precedents in granting patients the right to self-determination in the medical setting. By and large, judges have supported the rights of patients when competent patients have refused treatment. A class of significant exceptions is that of pregnant women, but as demonstrated in the case of Angela Carder, a court of appeals can override a trial court's decision. Such appeals are not usually sought because the performance of the cesarean section normally renders the case moot.

Judicial rulings have been sought even where there is no dispute between parties but cautious hospital administrators and risk managers require a court order to gain assurance that the proposed action is legal. An example is the case described earlier, in which the risk manager refused to allow a physician to perform an abortion on a comatose woman carrying a previable fetus. Adding to the caution was his refusal to seek the order on behalf of the hospital, requiring instead that the mother of the patient initiate the request for a judicial order.

The ruling by the District of Columbia Court of Appeals in *In re: A.C.* was unequivocal in its statement regarding coerced cesarean sections:

> We emphasize . . . that it would be an extraordinary case indeed in which a court might ever be justified in overriding the patient's wishes and authorizing a major surgical procedure such as a caesarean section. Throughout this opinion we have stressed that the patient's wishes must be followed in "virtually all cases," . . . unless there are "truly extraordinary or compelling reasons to override them". . . . Indeed, some may doubt that there could ever be a situation extraordinary or compelling enough to justify a massive intrusion into a person's body, such as a caesarean section, against that person's will.[23]

Pregnant Women Who Use Drugs and Alcohol

A strong movement has been joined by physicians, prosecutors, legislators, and others to coerce, detain, or incarcerate women who use drugs and abuse alcohol during pregnancy. This movement is not confined to the use of illegal substances, since alcohol is perfectly legal, but as is now well established, can cause problems for an infant as bad as or even worse than those caused by various illegal substances. There have been three prominent responses. The first is to identify women who are using drugs during pregnancy and to attempt some coercive action aimed at preventing them from continuing that behavior. The second is to do routine toxicology screening at birth and remove from their mothers' custody babies who test positive for drugs. The third is to seek criminal penalties for women whose babies are born with damage alleged to have been caused by their drug taking during pregnancy.

I contend that all three responses are ethically unacceptable. It is easier, of course, to say what is wrong with a particular social practice or public policy than it is to provide a recipe for a satisfactory solution. In addition to offering arguments in support of the position that coercive and punitive actions are ethically unacceptable, I can point to more enlightened approaches that are ethically superior and show some promise of ameliorating this terrible problem.

Under New York State law, a child who tests positive for drugs at birth is presumptive evidence of neglect or abuse by the mother. The practice has been to hold such children in the hospital until the baby's health status is stabilized and until a judgment is made about its custody. Until very recently, there were three alternatives. First, the baby could be released to the custody of the mother, but only if she succeeded in persuading social workers involved in the case that she was attempting to gain control over her drug use, usually by entering a drug treatment program. The problem has been that most drug treatment programs were not open to pregnant women. The second alternative was to release the baby to another family member, most often the maternal grandmother. Finally, placement in the foster care system operated under the auspices of New York City or private agencies has been the alternative chosen in about 50 percent of the cases. [24]

Commentators refer to the situation that prevails in many inner-city hospitals as the "drug wars." One observer noted that "Most hospital staff and child welfare workers are completely torn apart by the conflicts between their responsibilities to ensure such children's health and welfare and their concern for the mother's rights. In the harsh adversarial climate of the U.S. drug wars, . . . moral indignation often substitutes for provision of meaningful treatment and care for these infants and their mothers." [25]

Another commentator wrote:

> What I observe when looking at the drug war is that, with the help of the media, the "system" has confused who the enemy really is. The war on drugs has become the war on women who are poor, addicted, and in the minority. The weapon has become the "drug baby." The rights of women and their infants have been pitted against each other. [26]

In an interview published in the *New York Times*, [27] Dr. Jan Bays, Director of Child Abuse Programs at Emanuel Hospital in Portland, Oregon, was quoted as saying:

> We must up the ante to criminalize or impose reproductive controls on people who are out of control. Addiction is the most powerful force I have ever encountered. You have to use all the guns you have. . . . The nice thing about jail is that moms get good prenatal care, good nutrition and they're clean. . . . But we can't force people into treatment, even if they're in jail. She can go out and have more children. So, people are talking about sterilization and that gets into reproductive rights.

A sobering reply was offered by another physician, Dr. Ira J. Chasnoff, an expert on the effects of maternal drug ingestion on the developing fetus and subsequently born infants, who is president of the National Association for Perinatal Addiction Research and Education. The *New York Times* quoted Dr. Chasnoff as having said:

> This is a short-term, knee-jerk solution. The temperance movement is creating such a level of frustration that people are beginning to lash out at the group with the least defenses—women, especially the minority poor. . . .
> Criminalization of drug use by pregnant women won't accomplish anything in the long run. To develop punitive programs before we know the long-term effects of a mother's drug use on her children is ludicrous. . . .
> Furthermore, fear is not an effective deterrent because drug-using individuals are not reality-based and have strong denial mechanisms. They tell themselves they will never be caught.

These are the conflicting views of two medical experts on what approach to this problem is likely to be effective, as well as ethically acceptable. In addition, it is necessary to question Dr. Bays's contention that in jail, "moms get good prenatal care, good nutrition and they're clean." In general, prisons provide little prenatal or gynecological care. In one episode in California, a woman in prison "suffered severe abdominal cramping and bleeding for seventeen days without being allowed to receive treatment from an obstetrician; her son was born in an ambulance while she was being transported to an outside hospital, and lived only 2 hours."[28] In another California women's prison, a woman who was six months pregnant "suffered a miscarriage after she had been hemorrhaging and suffering abdominal pain for over three months. In spite of her critical condition she was only allowed to see an obstetrician/gynecologist on two occasions; as a result of the emergency nature of the miscarriage, she was also given a hysterectomy and is thus unable to have any more children."[29] These are only two such episodes, but they suggest that imprisoning women for the duration of their pregnancy is contraindicated from a medical and health standpoint, as well as raising serious questions about the ethical permissibility of the practice.

Although law and ethics are not identical, they overlap. Laws are often made (or repealed) for ethical reasons. Sometimes moral obligations are transformed into legal obligations. Let me turn now to an ethical analysis of this growing trend of legal coercion of pregnant women.

Ethical Analysis

It is useful to begin an ethical analysis by identifying points on which all can agree. There are at least two propositions on which universal agreement could probably be found. The first is that it is better for babies to be born healthy. Put another way, it is desirable for infants to be free from preventable diseases and developmental disabilities. This proposition is a value statement, but it is not meant to have any implications for those infants or children who are born with birth anomalies or diseases. It is perfectly consistent to hold that it is better for infants to be born healthy and sound than otherwise, and at the same time to maintain that infants born with disabilities deserve the same treatment and respect as healthy, able-bodied individuals.

The second proposition may appear somewhat less morally certain than the first, but I think it could also gain universal agreement. This is: Once a decision is made to carry a pregnancy to term, pregnant women have a *moral* obligation to act in ways likely to result in the birth of a sound, healthy infant.

This second proposition is a corollary to the first. If it is better for infants to be born healthy than unhealthy, and if behavior on the part of pregnant women can help to ensure that desirable outcome, then there is a moral obligation to act in ways most likely to bring about the desirable outcome.

Regrettably, these appear to be the only propositions on which most people agree. Views about what follows from those agreed-upon propositions differ sharply. The list of points about which there is debate or disagreement includes at least the following:

1. The pregnant women's moral obligation is directed at the fetus in utero.
2. The pregnant women's moral obligation should be transformed into a legal obligation.
3. The pregnant woman's obligation gives rise to a corresponding *right*, on the part of either the fetus in utero or the future child the fetus is likely to become.
4. The pregnant woman has an inviolable right to be free from coercive intrusions or invasions into her body, her liberty, or her privacy.
5. The fetus in utero has moral standing.
6. There exists an immoral act, which should be a crime, termed "fetal abuse."

None of these points follows directly from the first two propositions. Additional premises are needed to yield any of the conclusions embodied in the preceding six propositions, as well as other controversial statements about the rights and interests of pregnant women, embryos, fetuses, and future children.

Most ethical arguments contain a blend of an appeal to rights and obligations and an appeal to consequences. There are two leading traditions in moral philosophy—one in which moral rightness is a function of the best overall *consequences*, and a second in which the rights and obligations of the individuals involved in an ethical dispute are the central ethical features. Despite these two disparate perspectives, most actual ethical controversies draw on both traditions. Although both moral perspectives are respectable and relevant, I believe that a consequentialist approach to maternal–fetal issues is preferable. This is partly because it is less likely to make the fetus an enemy of the pregnant woman by pitting the alleged rights of one party against those of the other. But it is also because questions about rights in this situation remain unresolved and highly contentious, as the following analysis demonstrates.

How might an ethical analysis cast in terms of rights and obligations proceed? Once it is agreed that a pregnant women has a moral obligation to act in ways likely to result in the birth of a sound, healthy infant, the question arises: To whom is that obligation owed? The two leading candidates are (1) the fetus and (2) the child the fetus will become.

When the medical and ethical goal is the birth of a sound, healthy infant, the pregnant woman's obligation should be construed as an obligation to the future child, not to the fetus in utero. Construing the obligation in this way can be defended by a number of different considerations.

First, it is not for the sake of the fetus *while still a fetus* that a pregnant woman should take steps to promote its health and well-being. The moral obligation of pregnant women derives from the more fundamental value proposition that it is better for babies to be born healthy than unhealthy. The grounds for ascribing an obligation to pregnant women are concern for the health of infants and children. Good prenatal care is an important means to that end, but the object of moral concern is not the life or health of the fetus in utero.

Second, the moral status of the embryo and developing fetus is debatable. For those individuals whose religious beliefs accord the status of "personhood" to a human embryo, that entity has moral standing and requires protection of its life and well-being while still in utero. For all except right-to-life proponents, it is not present harm that matters, but future harm to the child the fetus will become. This is why the notion of "fetal abuse" is conceptually mistaken and ethically unsound. It implies that wrong or harm is being done to the fetus when, in fact, it is wrong or harm to the future child that constitutes the moral infraction. But since the future child does not yet exist, it is also a conceptual error to call pregnant women's behavior "child abuse." Neither the concept of "fetal abuse" nor that of "child abuse" can be accurately applied to situations in which pregnant women behave in ways that risk the health of their future child.

Like other moral obligations, the duty of pregnant women is contingent on a reasonable ability to comply. This stems from the philosophical precept that "ought implies can": that before people can be assigned moral obligations to act or refrain from acting in certain ways, it must be physically and psychologically possible for them to act in those ways. What constitutes a "reasonable ability to comply" with a moral obligation is often uncertain and open to dispute. In the case of pregnant women, a number of different constraints might limit their ability to comply with the obligation to act in ways that promote the health of their future child.

If the woman is a heroin addict, she may not have access to a treatment program. If she is an alcoholic, she may have tried—and failed—to combat her alcoholism. If she is a crack addict, her addiction might overpower her wish to do what is best for her future infant. Furthermore, at present, "most alcohol and drug treatment programs exclude pregnant women. And poor women have an especially hard time finding help. One survey of treatment programs in New York City found that 87 percent would not accept pregnant crack addicts on Medicaid."[30]

If a woman is a Jehovah's Witness, the strength of her religious belief may preclude her accepting a blood transfusion recommended for the well-being of the fetus. Another woman might refuse surgery out of religious convictions, an overwhelming fear of cutting, or after having experienced a relative's death from an anesthesia accident. If she has been advised by her obstetrician late in pregnancy not to have sexual intercourse, a woman may be unable to resist her husband's insistence out of fear of violence on his part. These or other circumstances can lead to legitimate questions about a pregnant woman's reasonable ability to carry out her obligation to promote the health of her future child. Applying the "ought implies can" maxim requires examining each individual circumstance to determine whether the obligation is one that the woman is capable of fulfilling.

In the absence of these sorts of extenuating circumstances, a pregnant woman's moral obligation is presumed to exist. A moral obligation to promote the health of the infant should *not*, however, be transformed into a legal obligation. Support for this proposition requires sound, persuasive arguments, and space does not permit a rehearsal of all the steps in those arguments. In brief, the different lines of supporting arguments are as follows:

To begin with, not everything that is immoral should also be made illegal. Many actions are morally wrong, yet are not subject to the force of law. To do so would convert our world into a completely legalistic one. Moreover, the intrusions into personal life that would be necessary for identifying and prosecuting people's failure to discharge their obligations would effectively eliminate the rights to privacy and to confidentiality that we so cherish.

A different argument contends that legal coercion of pregnant women is too strong a response to their behavior for at least three reasons: first, competent adults have a moral and legal right to refuse medical interventions that place them at risk; second, standards for taking away people's liberty by incarcerating them should be based on serious harms already inflicted or a high probability of serious future harm to another existing person; and third, the practice promotes social injustice because of the greater numbers of poor and minority women who will be suspected, reported, or indicted for the alleged "crime" of fetal abuse.

What about an ethical analysis that focuses on the consequences of actions? Does a consequentialist approach arrive at the same conclusion as an ethical analysis framed in the language of rights and obligations? This approach should take into account long-term as well as short-term consequences, as well as a broad array of outcomes beyond those that affect the individuals who are directly involved.

Among the potential consequences of legal coercion of pregnant women are the following:

First, there are consequences that flow from incarcerating pregnant women. The negative consequences for the health of the woman and the future child could be serious, as evidenced by the poor health care provided to pregnant women in corrections facilities. Among the additional negative consequences rarely mentioned is the effect on the other young children of mothers forcibly hospitalized or jailed during their pregnancy.

Second is the consequence of eroding the relationship of trust between physicians and their female patients. When physicians are transformed from advocates and allies of their patients into agents of the state—reporting them to government officials—the prospect of a good physician–patient relationship is greatly diminished. This argument was made in the case of Angela Carder in an *amicus curiae* brief submitted by the American Public Health Association (APHA). The brief stated:

> Rather than protecting the health of women and children, court-ordered caesareans erode the element of trust that permits a pregnant women to communicate to her physician—without fear of reprisal—all information relevant to her proper diagnosis and treatment. [31]

A third likely result is driving poor and disadvantaged women (those statistically more likely to be identified and reported as drug users) away from prenatal care, thus leading to even worse outcomes of pregnancy. This consequence was also mentioned in the APHA *amicus* brief: "An even more serious consequence of court-ordered intervention is that it drives women at high risk of complications during pregnancy and childbirth out of the health care system to avoid coerced treatment."[32]

Yet this predicted consequence has been challenged by Charles Condon, the solicitor in Charleston, South Carolina. Condon used child neglect charges against drug-using pregnant women in arrests beginning in 1989. By early February 1990, Condon said that arrests had almost stopped because the hospital was seeing fewer cocaine babies once drug-using women perceived they faced the risk of jail. And, Condon said, "there is no sign that addicted women are avoiding prenatal care or having their babies out in the woods to avoid arrest."[33] The prediction that drug-using women will be driven away from prenatal care is, like any other prediction, one that stands to be confirmed or denied by empirical evidence. Although the evidence cited from South Carolina is based on very few cases, it cannot be altogether discounted. Nor can it be used as a sound indication of what is likely to happen elsewhere.

Additional worries point to an impending slide down the slippery slope in the form of ever greater intrusions into the liberty and privacy of pregnant women, as well as an increased number of "medical indications" for coercion before, during, and after pregnancy.

There is one further consequence to consider: the "cost to society." It is common today to refer to the cost to society of having to provide medical treatment or institutional care for infants born with disabilities as a consequence of maternal behavior during pregnancy. Isn't this a legitimate issue? Isn't the cost to society a relevant concern?

My answer would be "yes" if health policy is based entirely on monetary costs and benefits but a firm "no" when ethical costs and benefits are considered. This is only one of many examples in which the prospect of financial burden is used to make an ethical argument. The question is whether it is "fair" to burden society with the cost of treatment for disabilities that might have been prevented. This line of reasoning cannot stand up to ethical scrutiny.

Those who confuse economics with ethics conclude that the right course of action is to coerce pregnant women, thereby preventing the need to spend money on care and treatment for possible birth defects. This approach pits the decision-making autonomy and the liberty of individuals against the economic interests of society. If a person acts in ways that may result in increased financial costs to society, then, according to this argument, removal of the individual's liberty in order to prevent incurring those financial costs is justifiable. But we cannot ignore the nonmonetary costs of eroding liberty and autonomy in a free society.

Whatever may be the proper response to women whose behavior during pregnancy risks the health of their future child, the ethical issues surrounding this problem should not be distorted by the importation of external factors like the cost to society. Although in fashioning social policy it is usually appropriate to include economic costs among the relevant factors, in an analysis of the moral rights and obligations of individuals, financial considerations remain external.

An especially troublesome feature of any consequentialist argument is the inevitable uncertainty that surrounds predictions of all sorts. There is often specific uncertainty in any particular medical situation about the likelihood and degree of harm to a fetus or future child. In addition, there is profound moral uncertainty regarding how high the likelihood must be, and how great the degree of harm must be, to warrant intrusions of privacy, limitation of liberty, coerced invasions of the body, or retrospective assignment of sanctions against the woman. Furthermore, when it comes to women who have poorer overall health and a lower nutritional status, it may be difficult to pinpoint a single cause of birth anomalies, low birth weight, or other afflictions an infant may have. In a society that places high value on individual privacy and liberty, it is a serious matter to contemplate the dismantling of legal protections for pregnant women in a climate of zealous concern for the fetus.

An ethical analysis in terms of consequences is less likely to render women and their fetuses adversaries than an approach that seeks to assign rights and duties. This is because the language of rights automatically pits individuals against one another and typically escalates moral claims in those rights that deserve legal backing. Yet rights cannot and should not be eliminated entirely from these ethical judgments. It is undeniable that forced medical treatment of pregnant women violates their moral and legal right to self-determination and informed consent or to informed refusal of treatment.

What about punishing women who used drugs during pregnancy after their babies are born? The arguments so far have primarily addressed the ethical issues pertaining to forced treatment of pregnant women and removal of their liberty during pregnancy. I have mentioned but not said much about postbirth sanctions against women whose drug or alcohol abuse during pregnancy, or whose failure to comply with medical recommendations, has been the likely cause of birth anomalies. Some writers have sharply distinguished the two situations, agreeing that forced interventions to protect offspring are a dubious public policy yet arguing that postbirth criminal or civil sanctions for maternal behavior that seriously injures offspring is justified.[34]

It is clear that a different set of arguments opposing legal coercion from those I have developed here would have to be used in cases where pregnant women's behavior has been the likely cause, or a contributory cause, of actual harm to the now existing child. Still, I think an argument against postbirth sanctions can be made. Three ethical considerations, taken together with facts already mentioned, prompt a rejection of postbirth legal action.

The first consideration recalls the "ought implies can" maxim. Drug- or alcohol-addicted women simply may not have it in their power to refrain from using these substances while pregnant, even if they are made aware of the likely imposition of legal penalties. Those who are informed and willing to enter a drug or alcohol treatment program are likely to find that treatment is not open to them. A determination that pregnant women should be held culpable and subjected to postbirth sanctions involves a decision to create a whole new category of criminal behavior, one whose perpetrators can only be women.

The second consideration points to the nature of the behavior. Although it is arguable whether the behavior of addicts and alcoholics is fully voluntary, it cannot be denied that when people use these substances they do so knowingly. But a large number of pregnant drug and alcohol abusers may still be ignorant of the consequences of their substance abuse on their future child. Women who use drugs and alcohol or engage in other medically noncompliant behavior during pregnancy do not normally do so with the *intention* of harming their future child. Nor are they *deliberately* seeking to inflict harm either on the fetus or on the child who will be born. While in the throes of addiction, addicts are not models of ethical behavior. A craving for the drug takes precedence over everything. Given the fact that addicts have been known to kill their lovers, mothers, or children for a fix, it is not surprising to find little regard for a fetus in utero.

The third consideration is the harm to infants likely to result from punishing their mothers. Separating children from their mothers is likely to cause psychological harm, since alternatives such as foster care are less than optimal. Mothers whose infants have been removed from their care are often reunited with them, further disrupting the continuity of parenting for a child. In addition, many women in this situation have older children at home, and they, too, will suffer from the separation if their mothers are imprisoned.

Taken together, these three ethical considerations can be used to reject the imposition of postbirth sanctions on women whose use of drugs or alcohol was a probable contribution to the harm of their infants. And although slippery slope arguments must always be used with caution, there is a danger of expanding the conditions under which postbirth sanctions could be imposed on women for their behavior during pregnancy: some alcohol ingestion as opposed to a lot; cigarette smoking; continuing to work in an environment with known occupational hazards, such as a hospital operating room or delivery room; and so on. A professor of health law who defends legal penalties asserts that "the desirability of post-birth sanctions should depend on the gains to children relative to the harms that might arise from such a policy."[35] As individuals and as a society, we are notoriously poor at predicting the benefits and harms that are likely to arise from public policies. I surmise that holding women criminally responsible for their behavior during pregnancy is likely to cause more harm than good. Proponents of legal sanctions suppose the reverse. Is there any empirical evidence that would lend weight to either prediction?

Ethical Conclusions

I have argued here that coercing pregnant women for the sake of the fetus cannot be ethically justified. The ethical principle known as the "harm principle" permits interference with an individual's liberty when there is a likelihood of harm to another person. Since there is no prospect of agreement on considering the fetus a person, and no scientific evidence that could ever settle that debate, the harm principle cannot be invoked as an ethical defense of interfering with the liberty of pregnant women. Despite the fact that some religious systems assert that personhood begins at conception, and despite the existence of some state laws and judicial holdings that a fetus is a person before the law, philosophical grounds for that conclusion are either wholly lacking or locked in unresolvable controversy.

Similarly, the harm principle cannot be used as a justification for limiting the liberty of pregnant women because of possible harm to future persons, that is, the person a present fetus is likely to become. A fetus might never become a person. It could die in utero. Or the pregnant woman might decide to have an abortion, a legal option still open to her in the United States. Finally, the fetus might be born and thus become an actual person, but without the harms, anomalies, or diseases that had been predicted.

One price of upholding individual liberty in a free society is the occurrence of some tragic birth defects or impaired children and the prospect of the death of a few viable fetuses. Although some tragedies are preventable, it is ethically unacceptable to seek to maximize prevention through legal coercion of the sort being proposed and already carried out with pregnant women. To seek to prevent some tragic illnesses and fetal deaths by erecting a system that pits physician against patient makes criminals out of women who risk the health of their future child, and to require women to be sedated and strapped down to undergo cesarean sections or blood transfusions is desperate and extreme. Analyzing these maternal–fetal conflicts using a consequentialist approach yields the conclusion that if legal coercion of pregnant women is ethically permitted and legally endorsed, a preponderance of *bad* consequences over good—both for women and for society generally—is likely to be the result.

I began with the premise that pregnant women have a *moral* obligation to refrain from behaving in ways that risk the health of their future child. None of the arguments against legal coercion serve to undermine that premise. However, pregnant women are not the only ones who bear that obligation. Duties fall on others as well.

Physicians have an obligation to educate and counsel their female patients who engage in behavior that risks the health of their future child. As physicians and healers, their obligation to their patients embraces the duty to strive to maintain a good physician–patient relationship, free from threats or coercion. Rather than act as policing agents of the state in seeking to detect and report drug use by pregnant women or positive screens in babies, doctors should be advocates of improved public policies in this area and help to work toward changing existing bad laws and introducing enlightened public health measures.

State, county, and city governments have an obligation to ensure access for all women to adequate prenatal care, as well as to treatment programs to combat drug and alcohol abuse. Governmental actions that could result in driving drug- and alcohol-using women away from prenatal care will most likely have the opposite effect from that intended: The more women who lack adequate prenatal care, the more premature, low-birth-weight infants and infants with disabilities are likely to be born.

As one example of a positive approach, an epidemiologist with the New York City Health Department described a new community-based case management program designed to provide a continuum of reproductive health care.[36] The program includes pregnancy testing, HIV testing and counseling, prenatal services, family planning services, and special services for teens, all to be carried out through field centers within each health district.

Other positive developments also appear to be underway. An article in the *New York Times* on September 19, 1991, reported on a plan intended to reverse the practice of routinely taking babies with a positive urine toxicology test away from their mothers. The article said that the plan was "aimed at keeping families together, encouraging drug treatment and saving millions of dollars in foster-care expenses."[37] A cynical view would conclude that it is the cost of paying for foster care for the 50,000 affected children that motivated this plan. But it should also be noted that the commissioner of the Human Resources Administration in New York City expressed a commitment to keeping families together and reuniting families that have been torn asunder for one reason or another. The new plan places certain conditions on the mothers, including continuing attendance at drug counseling sessions and classes on childrearing, and allowing a social worker to drop by periodically, sometimes unannounced, for a specified period of time. An encouraging note is the assertion by city officials that no children born exposed to drugs have suffered serious injury or death due to parental neglect or abuse since the new approach has been in use.[38]

A variety of efforts have been underway in California for a longer period of time. The state has taken a leadership role in developing programs for chemically dependent women and drug-affected children. As of the summer of 1991, there were 5 pilot projects funded by the state, with two more scheduled to begin shortly thereafter, and 27 demonstration projects funded by the federal Office for Substance Abuse Prevention.[39]

Another California project is located in Butte County, described as the "methamphetamine capital of the state."[40] We tend to think of addiction and perinatal drug problems as primarily inner-city phenomena, but the challenge of providing treatment services can be even greater in rural communities with limited professional and financial resources. The Butte County program developed a Perinatal Substance Project in late 1988, spurred on in part by threats by the district attorney to prosecute drug-using pregnant women. The project consists of case managers who coordinate a variety of services, including an in-home parent educator, social workers, and developmental specialists as needed, with additional support from other relevant local and regional agencies.

As these recent initiatives in New York City and California show, there *are* ethically acceptable alternatives to the three ethically unacceptable practices I have been discussing: detaining or otherwise coercing drug-using pregnant women; removing from their mothers' custody babies who test positive for drugs following a routine toxicology screen at birth; and seeking to impose criminal penalties for women whose babies are born with damage alleged to have been caused by their drug taking during pregnancy or simply discovered to have drugs in their system. Choosing the ethically acceptable alternatives over these repressive and punitive responses can serve to reestablish health care providers as advocates of their patients rather than casting them as enemies in the drug war.

Postscript in the Case of Angela Carder

The Court of Appeals in In Re: A.C. noted why it was issuing an opinion despite the apparent mootness of the case. The opinion stated that

> collateral consequences will flow from any decision we make in this appeal.
>   The personal representative of A.C.'s estate has filed an action separate from this appeal against the hospital, based on the events leading to the trial court's order in this case. In these circumstances, we adhere to our prior decisions refusing to

dismiss an appeal as moot when resolution of the legal issues might affect a separate action, actual or prospective, between the parties.[41]

That separate action resulted in another positive development to note in concluding this exploration of maternal–fetal conflict. Daniel and Nettie Stoner, Angela's parents, sought to ensure that tragedies similar to the one they experienced would not continue to occur. In addition to the challenge that led to the District of Columbia's Court of Appeals vacating the court order of June 16, 1987, which had required Angela to undergo a cesarean section, the American Civil Liberties Union filed a separate malpractice and civil rights suit against the George Washington University Medical Center for its treatment of Angela Carder and for its decision to involve the court. As a result of the settlement, the Medical Center developed a policy that now serves as a model for other hospitals to emulate.

Policy on Decision Making by Pregnant Patients

The policy adopted by George Washington University Medical Center places patient autonomy at the forefront. Management states: "We base our policies regarding decision making on this hospital's (and the medical profession's) strong commitment to respecting the autonomy of all patients with capacity.[42] The policy adds that respect for autonomy does not end when a patient refuses a course of action that physicians recommend. Moreover, the same ethical, legal, and medical standards that apply to nonpregnant patients also apply to the decision-making process of a pregnant patient.[43] The policy emphasizes the importance of counseling pregnant women and urges that when a pregnant patient's decision "appears unnecessarily to disserve her own or fetal welfare, great care should be taken to verify that her decision is both informed and authentic."[44] When such circumstances arise, the physician should seek to explore the reasons that lie behind the patient's decision.

Nevertheless, if all counseling and explorations with the patient fail, the ultimate decision is left to the woman. The policy states: "When a fully informed and competent pregnant patient persists in a decision which may disserve her own or fetal welfare, this hospital's policy is to accede to the pregnant patient's preference whenever possible."[45] The policy also addresses the issue of pregnant patients who are not capable of consenting to or refusing treatment in an informed fashion. In such cases, the document recognizes the authority of a surrogate for the patient, building in appropriate safeguards to protect the welfare of the patient and the fetus. The policy states that "the hospital will accede to a well-founded surrogate's decision whenever possible."[46]

Finally, and importantly, this model policy asserts that courts are an inappropriate forum for resolving ethical issues.[47] It endorses a strong commitment to keeping health-care decision making within the patient–physician relationship. The policy concludes with the statement that "it will rarely be appropriate to seek judicial intervention to assess or override a pregnant patient's decision."[48]

It is commendable for a hospital to issue a strong policy statement seemingly in conflict with its previous practices. A cynical view might hold that the change came about as a result of bad publicity and the reversal of the lower court's decision by the District of Columbia's Court of Appeals. But whatever factors were instrumental in the hospital's issuing this policy, its content represents a victory for the rights of pregnant women and a reiteration of the autonomy of all hospitalized patients. Both for its procedural elements and for its substantive ethical position, the document issued by the hospital is a model policy that all institutions would do well to emulate.

Notes

1. Deposition of Christine St. Andre, *Stoners* v. *G.W.U.*, Superior Court of the District of Columbia (1990), p. 69.
2. Ibid., pp. 176–79.
3. Deposition of Michael M. Barch, *Stoners* v. *G.W.U.*, pp. 53–64.
4. Deposition of William F. Minogue, pp. 34–35.
5. Ibid., p. 63.
6. *Amici curiae* brief, p. 5.
7. *In re: A.C.*, Appellant, Court of Appeals, No. 87-609, On Hearing en Banc, p. 1108.
8. Ibid., n. 2.
9. Ibid., p. 1109.
10. Ibid., p. 1121.
11. Ibid., p. 1122.
12. *Bonner v. Moran*, 75 U.S. App. D.C. 156, 157, 126 F.2d 121, 122 (1941).
13. *McFall v. Shimp*, 10 Pa. D. & C. 3d 90 (Allegheny County Ct. 1978).
14. Court of Appeals, p. 1123.
15. Examples and legal citations from Nancy K. Rhoden, "Informed Consent in Obstetrics: Some Special Problems," *Western New England Law Review*, vol. 9 (1987), pp. 67–88; and Martha A. Field, "Controlling the Woman to Protect the Fetus," *Law, Medicine & Health Care*, vol. 17 (1989), pp. 114–29.
16. Walter K. Meeker, Letter to the Editor, *New England Journal of Medicine*, vol. 317 (1987), p. 1224.
17. Cited in Rhoden, "Informed Consent," p. 82.
18. In the Matter of the Application of Jamaica Hospital for permission to transfuse blood into the person of Santiago X, Supreme Court, Special Term, Queens County, Part II, April 22, 1985.
19. Instances cited in Field, "Controlling the Woman," p. 115.
20. This occurred in the case of Pamela Monson Stewart in California.

21. ACOG Committee Opinion, "Patient Choice: Maternal–Fetal Conflict," no. 55, October 1987.

22. NOW Legal Defense and Education Fund, *Facts on Reproductive Rights: A Resource Manual*, "Court-Imposed Medical Treatment of Pregnant Women," Fact Sheet 15, New York: NOW LDEF (1989), p. 1.

23. *In re: A.C.*, p. 1142.

24. Ernest Drucker, "Children of War: The Criminalization of Motherhood," *The International Journal on Drug Policy*, vol. 1, issue 4 (1990), p. 1.

25. Ibid., p. 2.

26. Maria Leech, "Institutional Co-dependency and Recovery," *California Advocates for Pregnant Women Newsletter*, Issue 15 (Summer 1991), p. 1.

27. "Punishing Pregnant Addicts: Debate, Dismay, No Solution" *New York Times* (September 10, 1989), p. E 5.

28. NOW Legal Defense and Education Fund, "Facts on Reproductive Rights: A Resource Manual," Fact Sheet 13, "Punishing Women for Conduct During Pregnancy" New York: NOW LDEF (1989); cites Barry, "Quality of Prenatal Care for Incarcerated Women Challenged," *Youth Law News*, vol. 6, no. 1 (1985).

29. Ibid.

30. Tamar Lewin, "Drug Use in Pregnancy: New Issue for the Courts," *New York Times* (February 5, 1990), p. A14.

31. *In re: A.C.*, Court of Appeals, pp. 1131–32.

32. Ibid., p. 1132.

33. Lewin, "Drug Use in Pregnancy," P. A14.

34. See, e.g., John Robertson, "Reconciling Offspring and Maternal Interests During Pregnancy," in Sherrill Cohen and Nadine Taub (eds.), *Reproductive Laws for the 1990s* (Clifton, N.J.: Humana Press, 1989), pp. 259–74.

35. Ibid., p. 263.

36. Presented at a colloquium in the Department of Epidemiology and Social Medicine, Albert Einstein College of Medicine and Montefiore Medical Center, September 1991.

37. Joseph B. Treaster, "Plan Lets Addicted Mothers Take Their Newborns Home," *New York Times* (September 19, 1991), p. B1.

38. Ibid., p. B4.

39. California Advocates for Pregnant Women, *Newsletter*, issue 15 (Summer 1991), p. 2.

40. Ibid., p. 5.

41. Court of Appeals, pp. 1117–18.

42. Appendix A to Settlement Agreement, Policy on Decision-making by Pregnant Patients at The George Washington University Hospital, p. 1.

43. Ibid., p. 3.

44. Ibid., p. 5.

45. Ibid., p. 6.

46. Ibid., pp. 7–8.

47. Ibid., p. 9.

48. Ibid., p. 11.

5

Doctors versus Patients and Patients versus Patients

Are there situations in which physicians may justifiably view patients as their enemies? A patient may behave so badly that the doctor will dread any encounter. An especially manipulative patient can intrude into a doctor's personal life. For their part, some physicians refuse to care for patients who are noncompliant—who fail to take medication, who do not adhere to a prescribed diet, or who in other ways do not follow doctors' orders. Do these physicians then become enemies of their patients?

Some cardiologists will not take care of patients who smoke. Some obstetricians refuse to take care of pregnant women who consume alcohol. Such doctors take their patients' noncompliance as a personal affront. More enlightened physicians contend that patients who do not adhere to medically sound advice are their own worst enemies.

The traditional response of physicians to their noncompliant patients has been to blame the patient for these failures. More recently, a pro-patient response has turned the tables, seeking to blame the doctor for patients' failure to comply. In an effort to remove the victim-blaming stigma from patients, linguistic reformers have introduced the term "nonadherence" in place of "noncompliance." What usually happens when a neutral term is brought in to replace a pejorative label is that the new term eventually assumes the negative connotation of the discarded one. "Nonadherent" has already begun to take on the unsavory aura of "noncompliant."

There is little doubt that the problem of patients' failure to comply looms large in the eyes of physicians and other health professionals. The problem is revealed in the very definition of "compliance" that physicians use: "the extent to which a person's behavior coincides with medical or health advice."[1] From an ethical point of view, the question is whether noncompliance ought to place doctors and patients in an adversary relationship. Should the patient's failure to comply with the physician's advice be seen as an offense to the doctor? May the doctor "fire" the patient for such a perceived offense? Would that action thereby render the doctor an enemy of the patient? Is it appropriate to construe noncompliant patients as their own worst enemies?

A Difficult Patient

A problematic case, which involved several possible motives for physicians' rejection of a patient, attracted national attention in the media several years ago. The case involved Jeanie Joshua, a woman living in Ojai, California, whose physician sent her a letter stating that he was withdrawing as her physician. Ms. Joshua had kidney failure and required dialysis three times each week in order to survive. Dr. David Doner, the physician who was then caring for her, told the patient he could no longer care for her, since "she insisted on directing her own care to an extreme degree," and stated that serving as her physician was ruining his personal and professional lives. He was unable to leave town, either to go to professional meetings or on vacation with his family, since no other doctor would serve as his backup.

Ms. Joshua claimed that the reason the physician no longer wished to care for her was that she was in the process of suing other doctors in that community for malpractice, and this physician feared the same thing might happen to him. She contended that she was the victim of a "conspiracy" on the part of physicians to blackball her because she was suing another physician, Dr. Michael Fisher, who had previously supervised her on home dialysis. The filters on the home dialysis machine had malfunctioned, as a result of which red blood cells were destroyed and she required a transfusion. After that her condition declined, and she sued Dr. Fisher and the home nurse, as well as the manufacturer of the machinery. Ms. Joshua charged that the doctors who were asked to serve as backup physicians for Dr. Doner, her then current physician, were partners of Dr. Fisher, and they refused to treat her because of her pending suit against Dr. Fisher.

The physician and the patient in this case each had a different view of why the doctor refused to continue caring for her. Although we cannot know which account was more accurate, the case can still be analyzed from an ethical point of view. Dr. Doner was quoted as saying: "Our relationship has gotten very difficult. The emotional strain has gotten to the point where it's detracting from my freedom to grow professionally."[2] Kidney specialists in two different counties asserted their "right not to treat a patient they considered difficult and overly aggressive about directing her own medical care."[3] For her part, Jeanie Joshua admitted that she was "a big pain in the butt," then questioned: "What's that got to do with denying me life support?"[4] Patient and physician each had come to see the other as an enemy.

Dr. Doner apparently made the efforts required by law and professional ethics to find other physicians to care for his patient. He wrote to her, stating the reasons he could no longer care for her, and he supplied a list of other doctors she could consult. He made arrangements for her to see another physician, Dr. Kant Tucker, who told reporters that Ms. Joshua was "confrontational" when he spoke with her and "insisted on directing her medical care to an extreme degree."[5] Dr. Tucker told her, "As a physician, I am responsible for your health and you cannot make medical decisions by yourself." Dr. Tucker contended that Ms. Joshua rejected him as unsuitable to be her doctor. However, she claimed that when she told him about the litigation in progress, he "was obviously upset about it. . . . He said he didn't want to see me."[6]

The steps Dr. Doner took absolved him of "abandonment" in the strict legal interpretation of that charge. Furthermore, Jeanie Joshua was not in imminent danger of dying from kidney failure. Although none of the kidney specialists or the six dialysis centers in two counties would accept her as a patient, she could still receive dialysis by going to Los Angeles. As a result, three times each week, her husband had to drive her 150 miles to and from Ojai to Los Angeles so that she could receive the necessary life-sustaining treatment.

The reasons for a physician's refusal to care for a patient or a class of patients are critical in evaluating the ethical acceptability of such refusals. Also important for an ethical analysis are the surrounding circumstances: Can the patient's care be transferred to another physician? Is another physician or facility reasonably accessible to the patient? In the case of Jeanie Joshua, the answer to both questions was "no." No other doctor in the nearby area would accept the patient, and to receive dialysis she had to travel 300 miles, round trip, for the thrice-weekly treatments.

When may a physician who refuses to treat justly be viewed as an enemy of the patient? Physicians' rejection of a patient might arise out of self-interested motives, out of the perception that scarce medical resources require the physician to select only the most medically appropriate patients to receive those resources, or out of the doctor's prejudiced judgment of a patient's social worth. Only in the last situation can the physician correctly be termed an enemy of the patient. Where there is competition for limited resources, patients may become unwitting enemies of one another. As for the physician's self-interested motives, reasonable people disagree about the extent to which the obligation of doctors toward their patients requires a measure of self-sacrifice. In complex situations, all three factors may be operative, implicitly or explicitly. This complexity is illustrated in the following cases.

Three Noncompliant Patients

Ricardo, an 18-year-old boy with end-stage kidney failure, regularly misses his appointments for hemodialysis. He fails to adhere to the prescribed diet and eats bags of potato chips followed by quantities of Coca-Cola and beer. Periodically, after skipping a dialysis session on Friday, Ricardo is found unconscious by a friend or relative over the weekend and brought to the hospital emergency room. At other times, he shows up in the unit after missing a session, and because he requires immediate attention, the nurses have to ignore other patients in order to meet his urgent needs. Since his failure to show up regularly for dialysis is viewed as life-threatening, a psychiatric consultation is called to determine whether he is suicidal.

The evaluation judges Ricardo to be impulsive and defiant, yet lacking suicidal ideation. He fully understands his disease and the probability that one day he may not make it to the hospital after missing his regular dialysis appointment. The nurses on the unit ask whether Ricardo should be involuntarily committed to a psychiatric ward. The consulting psychiatrist replies that there is no justifiable basis for involuntary commitment. Although Ricardo lacks impulse control and recognizes the potential consequences of his behavior, he wants very much to continue living. Based on evidence from his own experience, he believes that he will be "rescued" whenever he gets into trouble. The staff experiences growing frustration and irritation. The nurses especially feel that they are neglecting other patients because of Ricardo's periodic need for emergency dialysis. They wonder if that is a reason to refuse to continue to provide dialysis for him. In the end, the decision is made to allow the situation to continue, despite the staff's anger and frustration and the belief of some professionals that the care of other patients is being unacceptably compromised.

Mr. O'Reilly has a 10-year history of alcohol abuse. He is from a middle-class background and has failed earlier nonmedical attempts to treat his alcoholism. Now he fears losing his job and expresses a sincere desire for effective medical treatment. He enters the hospital for a prolonged stay for detoxification, followed by drug treatment and counseling, and is then discharged. A month later, Mr. O'Reilly is admitted to the emergency department in a state of acute alcohol intoxication, vomiting blood. His physician claims that the patient has broken his "contract" and questions whether he is obligated to continue treating him. However, because the setting is the emergency department of a public hospital, the physician is precluded from rejecting the patient on two counts. The first is the patient's emergency need for treatment, which ethically compels the physician to treat him. The second is that a public hospital is a place of last resort, in which a refusal to treat would be socially and politically unsound.

Amanda McBride, a 23-year-old unmarried woman, is pregnant with her third child. She is receiving prenatal care in the outpatient clinic, and during an early visit she discloses to the obstetrician that the father of the baby, with whom she lives, is an intravenous drug abuser who has been tested and found to be HIV positive. Ms. McBride continues to have unprotected sex with the man and tells her obstetrician that he refuses to use condoms. She agrees to undergo a blood test to learn her HIV status, which turns out to be negative. The physician warns her that she is at risk of becoming HIV positive, and so is her baby, if she continues her present practices. Ms. McBride says she understands, but if she refuses to have sex with her man or insists on his using a condom, he will leave her.

The physician brings the case to the pediatric ethics rounds, expressing anger toward the patient and a feeling of helplessness. Although this physician stops short of suggesting that Amanda McBride should be institutionalized for the duration of her pregnancy in order to prevent her baby from being born with AIDS, others at the conference believe that coercion is justified in this situation. Of course, there are no ethical or legal grounds for forcibly detaining a person for the purpose of preventing acquisition of a disease, even a lethal one.

Explanations of Noncompliance

The three cases just described illustrate the wide range of behaviors that can fall under the heading of noncompliance. Failure to show up for scheduled therapy, inability to refrain from excessive alcohol consumption, and continuing a pattern of seemingly ordinary sexual behavior all count as noncompliance in certain circumstances. In such situations, tension is bound to occur between doctors and patients.

According to one assessment, "Noncompliance may be the most significant problem facing medical practice today. . . . Many reviews, involving hundreds of methodologically sound empiric studies, have repeatedly emphasized the dramatic magnitude and pervasiveness of the problem."[7] The results of these studies are revealing and, on the whole, rather discouraging. One study revealed that at least one-third of patients do not take medications as prescribed or take them in incorrect dosages or sequences. A review of several studies suggested that long-term compliance averages 50 percent, while noncompliance for short-term medication regimens has been as high as 92 percent.[8]

Peter Conrad, a sociologist, writes: "Throughout the literature published in the past two decades about patient compliance with medical regimens, especially drug regimens, the problem of noncompliance has attracted most of the attention."[9] There is even a publication with the title: *Journal of Compliance in Health Care.* It is not hard to see why this is a major topic of concern for health professionals. Since the behaviors recommended by physicians are designed to maintain or improve their patients' health—a goal patients themselves presumably desire—it is at least puzzling or frustrating to health care workers, and may even provoke their anger and hostility, when patients are noncompliant.

There are different types and degrees of noncompliance. Noncompliance may be a result of a patient's defiance or denial—either of the seriousness of the illness or of the need for treatment; it may also reflect a power struggle with the physician, forgetfulness by the patient, depression, or apathy.

Older, rather simplistic explanations have sought to explain non-compliance by attributing it to patients' uncooperative personalities, to their inability to understand a physician's instructions, or to irrationality. These are termed "victim-blaming explanations." They have recently been replaced by approaches "that regard the drug, the regimen, or the doctor–patient interaction as the source of the difficulty."[10]

Two other explanations have been offered by social scientists. "One theory locates the source of the problem in doctor–patient interactions—if physicians would give explicit instructions and clear feedback, or present advice in a friendly manner, patients would comply."[11] In contrast to the victim-blaming explanations, this alternative might be termed the "doctor-blaming explanation."

Still another explanation "suggests that patients' beliefs about their susceptibility to illness, the efficacy of the medications, and so forth are important in explaining compliance and noncompliance."[12] There is probably some truth in all of these suggested explanations, the older ones as well as the newer, more sophisticated accounts. As with other complex social and behavioral phenomena, it is likely that the search for a single, overarching explanation for patient noncompliance is a quest for the Holy Grail.

It has been well documented that a major cause of noncompliance is patients' adverse reactions to prescribed medication. One survey of 817 patients in a general practice found that 41 percent certainly or probably had a reaction to prescription drugs.[13] Studies of compliance with cancer regimens, in particular, reveal a high degree of noncompliance and termination by patients of these regimens because of non-life-threatening side effects such as nausea and vomiting.[14] If the physician's aim is to obtain better compliance, each variety of noncompliance calls for a different response. Sometimes the proper response is to alter the treatment regimen, rather than continue to seek compliance or lament the failure on the part of the patient, the physician, or both.

Among the wide range of factors that can affect patient compliance is the level of satisfaction with the patient–physician relationship. For example, studies have shown positive correlations between compliance and patient satisfaction with the visit, including perceptions of convenience and waiting time. Conversely, impersonality and brevity of the encounter have been shown to reduce patient compliance. Satisfaction and resulting compliance are greater when patients feel their expectations have been fulfilled, when the physician elicits and respects patients' concerns, when responsive information about patients' conditions and progress is provided, and when sincere concern and sympathy are shown.[15]

Given this complex picture, with its diverse causal factors, what can be said about the ethical issues raised by the problem of noncompliance? Is there only one ethical issue, or are there several, each dependent on the type, degree, or causes of noncompliance? Are there genuine ethical issues, or should physicians and other health professionals classify these concerns under the heading of "patient management problems?"

There are genuine ethical issues, and they center on the duties and obligations of physicians and the corresponding responsibilities of patients. But those obligations and responsibilities are not always clear. Do health professionals have the duty to care for their patients no matter how the patients respond? Are physicians who refuse to continue to treat blatantly noncompliant patients guilty of abandonment? Can patients' failure to comply with recommended treatment regimens legitimately disqualify them from further care? If patients refuse to uphold their part of the bargain, can they be penalized in return? Does labeling a patient noncompliant constitute yet another form of blaming the victim? Are physicians partly responsible for some instances of noncompliance by their patients? There is no simple answer to these questions, but with rare exceptions, the failure of patients to comply with a therapeutic regimen or to make medically recommended lifestyle changes does not release physicians from their obligation to continue to treat.

Conrad, the sociologist, observes that "The concept [of compliance] is developed from the doctor's perspective and conceived to solve the doctor's problem. The assumption that the doctor is the benevolent authority and patients are expected to comply is based on a consensual model of doctor–patient relations."[16] Embedded here is an implicit moral stance that "reproaches patients for their deviance and may devalue their claims to competency." This assessment concludes that "lack of compliance is one of the predominating characteristics by which doctors designate 'problem patients.'"[17]

Conrad points out the contrast when the situation is viewed from the patient's perspective. In his own study of people with epilepsy, using a fairly standard definition of compliance, 42 percent of the respondents were seen as noncompliant. Yet from the patient's perspective, the issue is not one of complying or not complying, but rather one of "self-regulation."

Noncompliance can thus be seen as an aspect of patients' struggles not to let an illness take undue control of their lives. "In this sense, it supports the person's desires for independence and autonomy, desires that align closely with the therapeutic goals of medical caregivers."[18] Yet having the same goals need not automatically lead to total compliance. An example is that of patients with diabetes who don't do exactly what their doctors prescribe: "While people with diabetes may have the same goals as their physicians, their method of achieving them may include altering their regimens to create a balance of caloric intake, exercise, and insulin."[19] Conrad proposes that "medical providers may want to examine noncompliance from a broader perspective. This is especially important with chronic illness, where an active role for patients may be necessary."[20]

But granting patients a more active role may not be enough if the aim is to bring about better outcomes. The numerous sociological studies have several implications for the obligations of physicians. Although these studies offer explanations for why patients fail to adhere to medical recommendations, they do not directly address the question of whether patients' noncompliance *ought* to alter their physicians' obligations toward them. That is one of the central ethical questions: *Should* physicians' duties or responsibilities be a function of their patients' compliance with medical advice? Pointing out that the standard version of compliance is yet another example of physician paternalism describes the problem more fully but does not provide an answer to the ethical question.

Doctors probably bear some responsibility for their patients' non-compliance, and they can assume a set of specific responsibilities that might succeed in improving compliance, at least in some patients. The ethical limits of attempting to improve compliance include refraining from resorting to coercive measures and communicating only truthful information. Other responsibilities include identifying patients' beliefs about health, disease, cures, and treatment; monitoring those health beliefs; appealing to patients' feelings and attitudes; being sensitive to patients' preferences; and being "certain that efforts to enhance compliance by modifying health beliefs are in the patient's best interests."[21] To accomplish all this would be a tall order. But there is evidence that efforts along these lines can be effective.

For example, one controlled study demonstrated that physicians who received training in patient education, related specifically to strategies for altering patients' beliefs and behaviors, spent more time educating their patients. Those patients later showed more knowledge and more appropriate beliefs about hypertension and its treatment than patients in the control group, whose physicians did not receive the special tutorials. In addition, the former patients were more compliant and had better blood pressure control as a result of becoming better educated about their disease and the treatment regimen.[22]

An ethical conclusion that can be drawn from this study is that physicians have a responsibility to take the time and trouble to educate their patients appropriately, and patients have a corresponding responsibility to learn something about their disease and its treatment. If doctors and patients alike accept these responsibilities and take them seriously, they will be more likely to end up as allies rather than as adversaries.

Unfortunately, in practice settings where physicians are paid on the basis of the number and complexity of the technological procedures they perform, the response to this proposal is likely to be "I don't get paid for talking to my patients." The true enemy here is reimbursement schemes that reward physicians only for doing medical procedures. Doctors' failure to adhere to their responsibility to educate and inform their patients can be traced at least partly to this factor and is likely to worsen in the future under "managed care" by armies of bureaucratic overseers.

Rights and Responsibilities

Is there a single conclusion to be drawn about the rights and responsibilities of both physicians and patients? At one extreme is the view that patients have all the rights and physicians all the responsibilities. Patients are sick, vulnerable, and wield less power, this view holds, so they are entitled to rights but are in no position to assume responsibilities. The doctor is obligated to treat, and to do so with caring and compassion, regardless of how ungrateful or obnoxious a patient may be. Doctors do not have to like their patients, but they have a duty to treat them in a way that is medically appropriate and respectful.

At the other extreme is the position that patients and physicians bear an equal burden of rights and responsibilities. As in other social arrangements, people are not entitled to rights unless they are prepared to exercise responsibilities. According to this view, patients can lose rights to which they would otherwise be entitled if they fail to act responsibly, that is, to comply with physicians' recommendations and to adhere to health-promoting behaviors. The situation might be likened to a contractual relationship between equals.

Somewhere between these extremes is the view that patients are vulnerable because they are sick and anxious, and they therefore have rights that they do not forfeit simply because they fail to comply with medical recommendations or adopt a healthy lifestyle. Nevertheless, patients do not have a right to demand anything or everything they might wish by way of medical treatment. Especially if a patient's behavior actually threatens the physician's life or safety, that patient does forfeit the right to treatment in that setting.

According to different so-called models of the physician–patient relationship, various answers are given to the ethical questions posed above about patient noncompliance. Some years ago, Robert Veatch, a major contributor to the field of bioethics, wrote an article entitled "Models of Ethical Medicine in a Revolutionary Age."[23] In that article, Veatch sketched four different models of the physician–patient relationship. He titled these the "priestly model," the "collegial model," the "engineering model," and the "contractual model."

The priestly model, as the name implies, conceptualizes the parties in the physician–patient relationship as unequal in power, authority, and prestige. It is the classical model of the paternalistic doctor who knows what's best for his patients. Even those unfamiliar with Veatch's other writings can surmise that he dismisses this model in favor of one that rejects paternalism and promotes the autonomy of the patient. Although the priestly model has the apparent virtue of following an ancient tradition guided by the principle "Benefit and do no harm to the patient," it has the undesirable feature of taking the locus of decision making away from the patient and placing it in the hands of the professional.[24]

The collegial model is also flawed, according to Veatch, because it is unrealistic. It conceptualizes the physician and the patient as colleagues or "pals." Veatch observes that "for the most part, we have to admit that ethnic, class, economic, and value differences make the assumption of common interest which is necessary for the collegial model to function . . . a mere pipedream."[25]

Veatch also rejects the third model—the engineering model—because it wrongly views the patient as the object of the physician's handbook or slide-rule calculations. This approach to the practice of medicine sees the physician as a "body mechanic." The engineering model conceptualizes the physician as "a plumber making repairs, connecting tubes and flushing out clogged systems, with no questions asked."[26]

It is the last of these four models—the contractual model—that Veatch holds to be the appropriate one for the physician–patient relationship. As the name implies, the relationship has, broadly speaking, the features of a contract, an agreement between parties rendered more or less equal by the assignment of obligations and responsibilities to each. The failure of either party to live up to these duties and responsibilities can be met with an appropriate response, spelled out explicitly or contained implicitly in the agreement or "contract." Veatch warns that the notion of contract should not be loaded with legalistic implications. But the "basic norms of freedom, dignity, truth-telling, promise-keeping, and justice are essential to a contractual relationship."[27] In contrast to the priestly model—that of benevolent paternalism—a different set of duties and obligations flows from the contractual model, which assumes and promotes maximal patient autonomy.

I am not alone in being skeptical about whether most actual physician–patient relationships can be made to possess these somewhat idealized features. The point of describing these models is to indicate how their differences give rise to different sets of duties and responsibilities. For example, the concept of the priestly model implies an attitude of forgiveness by the physician for the sins of the patient who fails to comply with a recommended regimen. The errant patient might "confess" the sin of noncompliance, to which the benevolent, fatherly physician responds with understanding and compassion. On the other hand, the paternalistic physician might scold the patient, in the manner of a father reacting to a naughty child or a judgmental priest to a penitent.

The engineering model appears almost amoral: We don't blame machinery for its failure to perform properly (although we may blame the manufacturer). We don't blame metal fatigue on steel beams that collapse. When our mechanical or electrical equipment wears out after a few years, we take it into the shop for repairs. That's the role of the physician in the engineering model: Fix the machine when it breaks down, but don't hold the machine itself responsible.

The contractual model comes closest to allowing for mutual obligations and responsibilities on the part of doctors and patients alike, but it remains defective in its unrealistic assumption of what is actual or even possible in the complex world of medical practice. It is also ethically problematic with respect to noncompliance. Since contracting parties have mutual obligations and responsibilities, the failure of either party to live up to the terms of the contract could render the contract null and void, as lawyers would say. But that is the wrong way to think about the doctor–patient relationship, and it yields a mistaken picture of physicians' responsibilities to their patients.

One reason why it is difficult to draw a simple, straightforward conclusion about health professionals' obligations to noncompliant patients is that the explanations for noncompliance are complex and varied. Moreover, many physicians adhere to the traditional view that it is their obligation to do what they sincerely believe is best for their patients. For those physicians, to respect a patient's autonomy can actually be unethical if it results in a bad medical outcome.

In contrast, radical autonomists hold that the right to self-determination is fundamental, in the medical setting as elsewhere. According to this ethical viewpoint, patients' rights trump physicians' obligations. A well-known example is the right of the competent patient to refuse life-sustaining medical treatment. Even a young, otherwise healthy patient, such as a Jehovah's Witness in need of a blood transfusion, has the moral and legal right to reject a physician's well-intentioned therapeutic efforts.

There is a critical ethical difference in situations where the patient is (1) unable to comply, (2) able but unwilling to comply, or (3) shares the physician's goals of treatment but chooses to alter the therapeutic regimen. In cases where the patient is psychologically unable to comply—for example, because of addiction to alcohol or tobacco, or because of a mental disability or disorder—the paternalistic model of the doctor–patient relationship can be justified. Such patients suffer from diminished autonomy; therefore, less responsibility should be ascribed to them. Here again, the "ought implies can" maxim comes into play. People who are psychologically or emotionally incapable of complying with a recommended regimen or making a change in lifestyle bear less responsibility for their failure. Recall the case of Mr. O'Reilly, who entered the hospital for detoxification. There is little doubt that he was motivated to stop the progression of his addiction, but also little doubt about the difficulty he faced after a 10-year bout with alcoholism. Was he truly unable to abstain following his hospital stay? Or was he simply unwilling?

Cases of unwillingness to comply are even more complicated than inability resulting from mental impairment or disorder. When a health professional suspects or learns that a patient is not adhering to a prescribed treatment regimen, the first obligation is to determine whether the patient is unable or unwilling. But even those who are able but unwilling to comply deserve continued care and compassion from their physicians. This is certainly true for cancer patients when the only available treatments often have side effects that the patient finds intolerable. However, if the patients' failure to comply is a function of dissatisfaction with the patient–physician encounter, as the studies mentioned earlier have shown, then there is a good deal physicians can do to alter their own behavior and, consequently, that of their patients.

When compliance can be improved by better physician–patient communication, by altering the patient's beliefs, or by increasing the patient's satisfaction with the doctor–patient relationship, physicians have as great an obligation to assist the patient in these ways as they do to make accurate diagnoses and prescribe the right course of treatment. Wherever the fault may lie, patients' noncompliance or lack of gratitude cannot justify refusal by physicians to continue to care for them.

What about patients' failure to make recommended lifestyle changes? Should these be viewed as different from medical regimens? Are lifestyle changes harder to comply with because of their broad and sweeping nature or because the harmful behavior has become entrenched over many years? In the case of Amanda McBride, the recommended changes would require her to alter her way of life and risk abandonment by her male partner because of his unwillingness to adopt safer sexual practices. If she lacks sufficient motivation or will to make these changes in the face of life-threatening consequences, there is probably little that physicians can do short of seeking to detain her for the duration of her pregnancy. That would be an ethically unacceptable intrusion into her privacy and deprivation of her liberty. To refuse to provide prenatal care can only increase the chances of a poor outcome for the baby.

Patients' refusal of life-preserving therapy constitutes the most extreme case of noncompliance. To the anesthesiologists described in an earlier chapter, refusal to treat Jehovah's Witnesses is well within the rights of physicians. This should serve as a reminder that people continue to disagree on the priority of different ethical principles: patients' right to self-determination versus physicians' obligation to benefit their patients and to do no harm. But it does not follow from the fact that people have conflicting subjective values that both sides are correct. Physicians do have an obligation to seek to do what is best for their patients. But they do not have a right to inflict unwanted treatment on an unwilling patient, abandoning those patients who fail to comply.

Noncompliance on the part of patients is only one reason why some physicians believe that their obligation to care for patients is lessened or even eliminated. As the case of Jeanie Joshua showed, physicians' own self-interest can also serve as a motive. Self-interested motives require some finer distinctions, however. Not all self-interested reasons are properly construed as selfish ones. A self-interested motive can be strictly personal yet possibly justifiable. Dr. Doner, in refusing to continue caring for Ms. Joshua, contended that being her doctor was ruining his personal and professional lives. Since no other physician in the community would care for her, he had to be perennially on call, which prevented him from attending professional meetings and taking vacations with his family. There is some merit in those contentions if they tell the whole story. In contrast, Ms. Joshua's allegation that the doctor's motive was really a fear of being sued would, if true, place his reason in the same category as that of the Georgia obstetricians who refused to accept as patients any lawyers, lawyers' wives, and employees of law firms.

Patients as Competitors for Scarce Resources

Sometimes physicians' refusal to treat certain patients stems from a desire to treat other patients instead, based on a real or perceived need to ration scarce or expensive medical resources. Deciding which patient should be placed in a critical care unit, in which beds or skilled personnel are a perennially scarce resource, forces physicians to reject some patients for the sake of others in need of that resource. By "scarce medical resources" I mean actual resources, such as organs for transplantation, beds in critical care units, and specialized equipment in hospitals. This is a role forced on those physicians responsible for allocating resources in critical care settings such as cardiac care and neonatal intensive care units.

When a doctor rejects a patient for a bed in the intensive care unit or as a candidate for an organ transplant, it is easy to think of the doctor as an enemy of the patient. But only when the physician's denial of a medical resource is based on a judgment of the patient's social worth can the physician properly be thought of as the enemy. In the case of genuinely scarce resources, it is patients themselves who become competitors of one another.

At a pediatric conference, the topic of discussion was ethically proper allocation of the high-technology treatment known as extracorporeal membrane oxygenation (ECMO). The treatment is necessary for survival in some children, but supplies are limited and not all who need the therapy are able to receive it. In adult intensive care units, the practice known as "triage" is a common method of allocating the scarce resource. Patients who are doing well are moved out of the unit to make room for those in greater need of the beds, and some patients deemed hopeless or near death may also be transferred in order to admit more recently arrived patients. But in pediatric units, it is standard practice to treat children in the order in which they appear for intensive or critical care, and to keep treating them even if children who arrive later can benefit more from therapy. However, some children who might derive benefit from ECMO are not even started on it. This is because of the medical judgment that another child is likely to show up soon who will derive much greater benefit.

One physician at the conference argued that to use the method of triage in pediatric intensive care units would make patients enemies of one other. It would be unacceptable, he claimed, to pit one patient against another and to create a situation where parents would view other children receiving care as rivals of their own child. But it is ethically arguable which system is more just, and whether the prospect of patients becoming enemies in competition for resources is eliminated by one method any more than the other.

I was surprised to learn that some children who might benefit from ECMO were denied access even if therapy could be started. Keeping a treatment slot open for a future patient, one not yet on the scene, seemed to be a questionable practice. Yet to insiders in the medical world, that practice is readily accepted. It is also more widespread than one might think. Not only in pediatric units but also in adult critical care units, beds are deliberately kept unoccupied for future patients. Whether this practice can be ethically justified requires an analysis of each particular case.

In one case on an adult intensive care unit, a 77-year-old man was admitted to the emergency room of the hospital with acute chest pains. He came from a nursing home, was slightly demented, and was a double amputee. Although information from the nursing home revealed that he had a son and a daughter, they did not accompany him, nor did they visit at any time during his hospital stay. Repeated efforts were made to admit this patient to the cardiac care unit (CCU), and the request was turned down each time. This was done in spite of the fact that there was an empty bed in the CCU.

The director of the CCU gave these reasons for not admitting this patient: "Because of his age, dementia, and medical history, he's not a good candidate for the CCU." When asked about the empty bed, the director stated that it was standard practice to keep a bed in the unit open. "You never know who's going to show up later tonight."

Is it ethically justified to keep a bed open for this reason? The case would be clearer if there were another actual patient in competition with this patient for the last bed in the CCU. On whatever grounds a choice might be made between two actual patients in competition for a scarce resource, the basis for denying a bed to an actual patient when the competition is an unknown patient who has not yet arrived remains unclear to me and seems ethically questionable. Yet it is defended by insiders who argue that their experience assures them that the best use of this limited resource can be made by holding a bed open. The underlying principle appears to be utilitarian: Seek to maximize the benefits of a scarce medical resource for all who are likely to need it.

One medical student commented, "If the patient had had two legs instead of stumps, he would have been admitted." Another observed that the man's son and daughter didn't visit, but that when a patient's family is hovering about and is very much involved, that often influences a decision about intensive care unit placement. If true, these observations show that it is rarely patients' prognosis alone that determines whether they will receive a scarce resource. Factors unrelated to the patient's medical prognosis influence physicians responsible for allocating hospital resources such as intensive care unit beds.

In a second case, the patient was a 53-year-old man who had suffered weakness in his legs for three weeks. He was brought to the hospital in an impaired mental state by the emergency medical service. The man was unkempt, dirty, and lethargic. It emerged that he had a history of more than 30 years as an alcoholic, consuming more than a pint a day. On physical examination he was found to have signs of liver disease. His leg weakness was attributed to complications of alcoholism; his mental status improved after he was treated with antibiotics.

After several days in which the patient's overall status did not improve, he was presented to the intensive care unit. The director of the unit refused to admit him on the basis of a poor prognosis. At that point the man's kidneys were failing, so a renal specialist was called in. The consultant thought that this condition might be reversible and suggested that he be given a full evaluation, which required placement in the intensive care unit. The patient was presented to the unit for the second time and again was refused a bed. A second visit from the consultant yielded the judgment that the patient was not suffering renal failure as a result of alcoholism and liver disease, and he was presented to the intensive care unit for the third time. The director again refused, saying, "We have to keep a bed open in the ICU for citizens who might come in needing a bed." The patient remained on a regular hospital ward and continued to deteriorate. He was intubated to assist his breathing but died a few days later.

Intensive care unit beds are a truly limited resource and should be allocated wisely on the basis of sound criteria. A poor prognosis can be a sound justification for refusing a patient a bed. But in order to be ethically justified, that refusal should be based on a medically competent and honest evaluation of the particular patient's prognosis. In the case of the alcoholic patient, the specialist stated that the patient had reversible kidney failure and could benefit by placement in the intensive care unit. The director's refusal to admit the man might have been based on his generally poor prognosis as an alcoholic of more than 30 years' duration. It was allegedly based on the judgment that a "more deserving" patient would show up later. "More deserving" could mean "having a better medical prognosis," but as a comment overheard on the wards implied: "He would have been admitted if he was 'more of a citizen.'" If a patient who had an analogous medical condition but was not an alcoholic had been presented to the intensive car unit with an open bed, his admission would have proceeded without a hitch.

A refusal to treat one patient on the grounds that another patient competing for a scarce resource might come along later should be scrutinized in an ethical analysis of the case. However prejudicial and infused with the physician's own values such a refusal may be, the grounds for refusal are still patient-centered. The choice of one patient over another as more deserving of a scarce medical resource represents a value choice, with some values being more ethically justifiable than others. But these are situations in which one patient or another will get the medical treatment. It is an entirely different matter when refusal to treat extends to an entire class of patients. This is what is now happening with intravenous drug users who develop heart disease.

Refusing Heart Surgery to Intravenous Drug Users

Most people act from motives that are mixed or uncertain, and physicians are no exception. Mixed motives can be detected in surgeons' refusal to operate on addicts who have reinfected their heart valves. In addition to appropriate medical reasons that stem from assessing the patients' prognoses, these motives can include the surgeons' perception of scarce medical resources, their fear of acquiring HIV infection, and judgments of the social worth of these patients.

The heart valves of intravenous drug users become infected as a result of repeated or long-term needle use. The valves can be replaced surgically and the infections can be treated medically. But that treatment is costly; involves hospitalization and the use of valuable time and skills of busy cardiothoracic surgeons; and often requires stays in the intensive care unit, where beds are always a scarce resource. Surgeons, both individually and as a group, have begun to question their obligation to these patients. An added factor is the prevalence of AIDS among intravenous drug users. With approximately 50 percent testing positive for the HIV antibody, surgeons are concerned about their own risk of acquiring AIDS while engaged in bloody cardiothoracic surgery.

A cardiothoracic surgeon presented the following case at a meeting of the hospital ethics committee.

The patient was Derek Williams, a 38-year-old Vietnam veteran. Mr. Williams held a job, and had a wife and one child. He had been an intravenous drug user for years, with heroin as his drug of choice. His diagnosis was an infected aortic heart valve.

Williams had been a regular patient as a Veterans Administration (VA) hospital in another borough of the city and had had one valve replaced in March 1988. Following the surgery he used heroin again, and his prosthetic valve became infected in August of that year. He went back to the hospital where he had been treated, and an abscess around the valve was found. The patient was given a one-month course of antibiotics and was discharged. Again he used heroin, went back to the hospital, and allegedly was told: "You had your chance." Williams claims he was told to try one or two (named) private hospitals. When later questioned, personnel at the VA hospital denied having made that suggestion.

The surgeon presenting the case acknowledged some medical uncertainty about whether this was a failed cure of the past infection or a reinfection from repeated drug use. He also wondered whether that mattered with respect to his obligation to treat the patient.

Derek Williams did not now have heart failure, nor had he developed other complications of bacterial endocarditis. But clinical assessment indicated the high likelihood that surgery would be necessary in six weeks, both to replace the newly infected valve and to repair the cavity resulting from the abscess.

Surgeons at the VA hospital where Williams had been seen were reported to have said: "We only operate once. Patients like this are not given a second chance. Although it's not written down anywhere, this is the unofficial policy. Resources are scarce, and we don't do a second valve replacement when the patient has reinfected a valve by shooting up."

This remark illustrates the widespread tendency to confuse genuinely scarce medical resources with those that are expensive. Physicians and the general public now routinely refer to all medical resources as scarce, when instead many are costly rather than truly limited. This confusion has led to calls for rationing medical care in ways that are ethically suspect (whether physicians have an obligation to refuse some patients because of costly but not genuinely scarce resources is taken up in Chapter 8). Yet to refuse to reoperate on drug addicts who have "caused" their own disease because of a higher obligation "to society" appears to be a more noble motive than the belief that this class of patients doesn't deserve the surgeon's time and effort. Even if it is true that the cost of operating on these patients is ultimately borne by society, it is not the surgeon's proper role to act as a fiscal gatekeeper.

The surgeon who presented the case to the ethics committee noted three factors that make a cure of the patient's immediate condition unlikely: the infection, the prosthesis, and the abscess. Does the likelihood of a negative outcome even if treatment is administered provide a good reason for denying the patient surgery? Refusal to operate could be justified under either of two conditions: (1) if the surgery itself posed an inordinately high chance of mortality or (2) if the operation would be futile, that is, if the patient's overall condition and prognosis would remain the same even after surgery. However, the surgeon reported that in a small series of patients having the three risk factors (infection, prosthesis, abscess), most came through the surgery reasonably well. Therefore, the operation could not reasonably be denied on medical grounds to be consistent with standards widely accepted in surgical practice.

Now, however, the surgeon mentioned an additional complicating factor: Mr. Williams is HIV positive. He has no clinical signs of AIDS or ARC. Yet according to one medical view, repeated infections may be the first sign of immunodeficiency. Based on the experience of a San Francisco group, 30 percent of HIV-positive patients with infected heart valves developed full-blown AIDS in 5 years and 50 percent in 10 years. Are these prognostic implications of HIV disease reasons for denying the patient surgery? Unless a patient is septic, with systemic disease caused by the presence of micro-organisms or their toxins in the blood or tissues, the prognosis from HIV disease is not a medical contraindication for surgery.

So, the surgeon concluded, the real question isn't Derek Williams's HIV infection, but rather his heroin addiction. The surgeon asked: Can permanent addiction be considered a fatal condition? Does a person who has had endocarditis have a fatal condition? This surgeon's experience, going back to 1978, was that all patients who were addicted and had endocarditis—with one exception—had died. More generally, the clinical experience of cardiothoracic surgeons confirms that the recalcitrant addict who has been operated on once ends up dying of his disease, *whether or not* another valve replacement is needed. So the prognostic indications for surgery based on Derek Williams's "disease" of intravenous drug addiction differ considerably from the prognostic indications stemming from his HIV infection.

The position of this surgeon had once been that performing these operations was not medically contraindicated. Furthermore, he had refused to view drug-addicted patients as guilty of causing their own disease and, therefore, as undeserving of treatment. But he was being worn down by his surgical colleagues. Increasingly, he stood alone as one who believed there was an obligation to treat.

To sort out the obligations of doctors toward intravenous drug users whose behavior has caused their heart valves to become infected requires some comparisons with other cases. Consider the case of a patient who is not a drug addict and needs surgery. Suppose that the patient has incurable cancer that will cause her death in the same period of time that the drug addict's addiction will cause his death. The prognosis in terms of the number of years of survival would be the same. Do the two patients deserve different treatment? One difference is the addict's so-called voluntary contribution to his own disease. Is that consideration ethically relevant from the standpoint of whether an obligation exists to perform surgery? A second potential difference is the possibility that the intravenous drug user might change his behavior, thus leading to a quite different prognosis.

Some argue that patients who have caused their own disease are not entitled to a share of scarce or expensive resources, including the time required for highly skilled surgeons to perform the operation, the care delivered by nursing staff, and a bed in the intensive care unit. Others claim that patients' past health-risking behavior should not be taken into account in considering whether to treat them, but that the likelihood of their continuing that pattern of behavior in the future should be a factor. Still others in this debate claim that patients' contribution to their own disease is not a morally relevant factor and should never be used in determining whether treatment should be administered.

There appears to be a nationwide trend: Cardiac surgeons are not doing repeat valve operations. Many avoid operating on such patients following the first infection. According to the surgeon who presented the case of Derek Williams, some hospitals in Miami have agreed to share the addicts: Each will do one valve replacement, and no hospital will do a second operation. Most recently, a medical student at an ethics conference reported a case in which surgeons were reluctant to perform the first valve replacement, and delayed the operation to the point where the patient deteriorated and was no longer a suitable candidate for surgery.

Can these surgeons truly be considered enemies of the patients who come to them? Or are they justified in refusing to replace heart valves—at least after the first replacement—in patients who went back to drug use after they were successfully treated? It is tempting to think of these patients as enemies of themselves. That view is held by those who argue that patients who cause their own disease deserve less of society's resources than people who become sick through no fault of their own. I contend that there is no obvious and nondiscriminatory way to draw a line between people who contribute to their health or illness and those who do not. Prejudice is typically greatest against individuals whose behavior or lifestyle is socially undesirable. Patients from the lower economic classes who are perceived as a drain on society, and those who are addicted to drugs or alcohol, are much more likely to be seen as undeserving of treatment than privately insured patients who are obese, addicted to nicotine, have a high cholesterol count, and never exercise. It is unjust to single out the former group of noncompliant patients, denying them treatment or admission to an intensive care unit, while the latter group are seen as "citizens" worthy of surgeons' time and the hospital's resources.

Yet an ethical question remains: Is there a point at which the obligations of physicians to patients cease? I have argued that that point is not reached either in the case of noncompliant patients or in the case of those who contribute in some way to their own illness. However, another category of patients might be potential candidates: those who are perceived to pose a threat of some sort to physicians. Whether a threat of harm is truly significant or is exaggerated remains to be explored. But situations do arise in which physicians claim that their refusal to treat is justified by unacceptable risks to themselves, situations in which they view the patient as an enemy.

Notes

1. R. Brian Haynes, David W. Taylor, and David L. Sackett (eds.), *Compliance in Health Care* (Baltimore: Johns Hopkins University Press, 1979), p. 1.

2. Jesse Katz, "Rights Tested as Kidney Patient Fights to Survive," *Los Angeles Times* (October 27, 1988), Part IX, p. 1.

3. Ibid.

4. Robert Reinhold, "When Doctors Shun Difficult Patients," *New York Times*, National Edition (November 11, 1988), p. A16.

5. Katz, "Rights Tested."

6. Ibid.

7. Stephen A. Eraker, John P. Kirscht, and Marshall H. Becker, "Understanding and Improving Patient Compliance, *Annals of Internal Medicine*, vol. 100 (1984), p. 258.

8. Ibid.

9. Peter Conrad, "The Noncompliant Patient in Search of Autonomy," *Hastings Center Report*, vol. 17 (1987), p. 15.

10. Ibid.

11. Ibid., p. 16.

12. Ibid.

13. Eraker et al., "Patient Compliance," p. 259.

14. Carol Lewis, Martha S. Linet, and Martin D. Abeloff, "Compliance with Cancer Therapy by Patients and Physicians," *The American Journal of Medicine*, vol. 74 (1983), p. 674.

15. Eraker et al., "Patient Compliance," pp. 260–61.

16. Conrad, "The Noncompliant Patient," p. 15.

17. Ibid., p. 16.

18. Ibid.

19. Ibid.

20. Ibid., p. 17.

21. Eraker et al., "Patient Compliance," p. 260.

22. Ibid., pp. 261–62.

23. Robert M. Veatch, "Models of Ethical Medicine in a Revolutionary Age," reprinted in Samuel Gorovitz, Ruth Macklin, Andrew L. Jameton, John M. O'Connor, and Susan Sherwin (eds.), *Moral Problems in Medicine*, 2nd ed. (Englewood Cliffs, N.J.: Prentice-Hall, 1983), pp. 78–82.

24. Ibid., p. 80.

25. Ibid., p. 81.

26. Ibid., p. 79.

27. Ibid., p. 82.

6

Doctors Who Feel Threatened by Patients

Medical students new to the wards were discussing a patient who appeared to be a candidate for surgery. "So after we did the medical workup, we called for a surgical consultation. The surgical resident came by and evaluated the patient. He said the guy needed the procedure," one student reported.

"What was the ethical problem in the case?" I asked.

"Well, the patient was scheduled for surgery the next day, but in the morning when he was supposed to be prepped, the surgery had been canceled."

"What was the reason?"

"The surgeons said there were more urgent cases. So they blew this patient off the list."

"Isn't that a good reason to postpone surgery for one patient? Other patients have a more urgent need, so there's a problem of scarce resources."

"Yes, if that was really true," the medical student replied. "But the patient was an intravenous drug user, and we think that's why the surgeon blew him off the list. There's a good chance he's infected with the AIDS virus. We know from other cases that surgeons try their hardest not to operate on patients they suspect of having HIV infection."

Are doctors ever justified in refusing to care for patients? Must physicians undergo risks not only to their health but even to their very lives? Since the AIDS epidemic began in the early 1980s, these questions have been asked with persistent urgency. In medical practice today, patients with HIV infection may pose some risk to physicians and other health workers who care for them. The risk is felt most acutely in specialty fields such as surgery, where health professionals come into direct contact with patients' blood, sometimes in large amounts.

Some physicians are reluctant to care for any AIDS patients, while others refuse only to perform specific procedures that carry an elevated risk of transmission of the AIDS virus. This is an issue of great moment in many hospitals and in particular regions of the country. It requires a careful analysis of the facts surrounding HIV transmission. Although physicians are at risk of acquiring other serious infectious diseases, such as hepatitis or tuberculosis, from their infected patients, it is virtually unheard of for doctors to refuse to care for patients with these diseases out of fear of acquiring the disease. AIDS has prompted a different response. Whether physicians who refuse to care for AIDS patients should be termed enemies of those patients is more than a semantic quibble.

Does the risk of acquiring a lethal disease from a patient justify a physician's or surgeon's refusal to treat? Most attention has focused on surgeons who are reluctant or who refuse to operate on patients with HIV infection, as in chest surgery, which is typically very bloody and carries a risk of transmission through accidental punctures of protective equipment. Another, less frequently discussed example is that of patients who are in need of emergency treatment such as mouth-to-mouth resuscitation. How high must the probability of harm be to warrant a health professional's refusal to care for a patient with AIDS? How should doctors and nurses act in the face of uncertainty about the risk of transmission?

It would surely be a mistake to conceive of patients with an incurable infectious disease as enemies of health care workers. Physicians and nurses have a general obligation to care for HIV-infected patients, as they do for any others in need of treatment. Some surgeons have called for screening all of patients admitted to the hospital for HIV infection; others have insisted on testing individual patients and have refused to perform elective surgery on those who test positive. More than the general public, health professionals should be completely informed about the large body of evidence demonstrating that HIV infection cannot be transmitted by casual contact. They should also be aware of preventive measures and should keep protective equipment readily available. The following case was brought to a meeting of a hospital ethics committee by physicians who hoped to obtain some guidance for the future.

A patient with a diagnosis of AIDS was admitted to the medical intensive care unit with *pneumocystis carinii* pneumonia. She was seven months pregnant at the time. The fetus was known to have endured several episodes of distress in recent weeks. Unexpectedly, the patient went into labor while in the intensive care unit. No one from the obstetric or pediatric service was present. The infant emerged, blue but moving. A quick search revealed that no protective equipment for neonatal resuscitation was present in the adult intensive care unit. Two physicians and a nurse who were in the unit at the time hesitated momentarily. One ran to the neonatal unit to get the equipment. Another began mouth-to-mouth resuscitation. Efforts to resuscitate the premature infant proved futile. The mother died several days later. The intensive care unit staff wondered whether they were obligated to perform mouth-to-mouth resuscitation in a case such as this.

In the era of AIDS, are physicians and other health professionals obligated to perform mouth-to-mouth resuscitation? Since CPR is a procedure typically carried out in an emergency, a "no" answer would have sweeping implications for the way emergency medical care is delivered, seriously undermining the traditional duty of health care workers to their patients. A "yes" answer would give rise to an objection stemming from the ethically sound premise that physicians are not required to submit to unreasonable risks. As posed, the question is poorly formulated; the issue can be addressed properly only by breaking the inquiry into several parts.

Are physicians and health care workers obligated to perform mouth-to-mouth resuscitation on individuals known to be HIV positive? On individuals whose HIV status is not known but who are in a hospital or catchment area where the prevalence of HIV infection is high? And how about patients whose HIV status is not known but who are in a hospital or catchment area where the prevalence of HIV is low? Even if the question is broken down into these subcategories, it is hard to give a general answer. What if the patient with an unknown HIV status is a young man brought to the emergency room with bloody saliva? What if the patient is an infant, abandoned in the hospital after birth by an intravenous drug-using mother? What about the risk to emergency medicine personnel who are pregnant? What if the patient in need of emergency resuscitation is already terminally ill?

These questions have special urgency because of the overwhelming ethical consensus that in an emergency, health professionals are obligated to treat to the best of their ability. A patient in need of resuscitation, whether HIV infected or not, is clearly in an emergency situation. To make exceptions to the traditional rule that, in an emergency, life-preserving treatment must be administered is to open the door to discrimination against people with AIDS, people suspected of having HIV infection, and even patients about whom there are no grounds for suspicion other than that they have entered a hospital in a particular geographic location. Yet there may be good ethical reasons for seeking to draw the line beyond which emergency treatment is not obligatory while still providing safeguards to prevent discrimination against whole classes of patients.

Current data regarding HIV transmission provide little to go on. It can be safely concluded from the available evidence that the risk of seroconversion following exposure to saliva is low. But can it be concluded with the same confidence that the risk of seroconversion following emergency mouth-to-mouth resuscitation is low? HIV has been isolated from saliva far less frequently than from the blood of HIV-positive individuals.

Research suggests that human saliva contains substances that prevent HIV infection, and that the fragile virus is inactivated in the acidic environment of the stomach. These results may help to explain why no epidemiologic evidence exists to suggest that contact with potentially contaminated saliva has resulted in seroconversion. But they provide little to substantiate the supposition that emergency resuscitation on HIV-infected individuals poses a low risk to the person performing the resuscitation, and they are of no relevance to the case of resuscitation of a newborn covered with blood from its HIV-infected mother.

A number of professional groups have issued statements or position papers regarding the obligation of their members to care for HIV-infected patients. These statements differ in the degree of obligation they impose on health professionals. None directly addresses the issue of mouth-to-mouth resuscitation.

The Council on Ethical and Judicial Affairs of the American Medical Association issued a report in 1987 that addressed the general issue of refusal to care for patients with AIDS. The report states: "A physician may not ethically refuse to treat a patient whose condition is within the physician's current realm of competence solely because the patient is seropositive. Persons who are seropositive should not be subjected to discrimination based on fear or prejudice."[1] Is refusal to administer mouth-to-mouth resuscitation to a seropositive patient "discrimination based on fear or prejudice"? Or is it justifiable avoidance of excessive risk to the physician? The report does not say.

The report does, however, recall Principle VI of the 1980 Principles of Medical Ethics, which states: "A physician shall in the provision of appropriate patient care, except in emergencies, be free to choose whom to serve." It is evident that cardiopulmonary resuscitation is an emergency procedure. This statement implies, then, that physicians are ethically required to perform mouth-to-mouth resuscitation. But the obligation is not made explicit.

The Committee on Ethics of the American Nurses' Association (ANA) in November 1986 issued its "Statement Regarding Risk V. Responsibility in Providing Nursing Care." Drafted to include infectious diseases other than AIDS, such as typhoid, tuberculosis, hepatitis B, and influenza, the statement asserts that "accepting personal risk which exceeds the limits of duty is not morally obligatory; it is a moral option."[2] The ANA report states that "in most instances, it would be considered morally obligatory for a nurse to give care to an AIDS patient."[3] However, the only example provided as an exception is a situation in which the nurse's own immune system is weakened, thereby greatly increasing the risk of acquiring an infection.

Mouth-to-mouth resuscitation is not specifically addressed. But the ANA statement does contain a provision that might apply to CPR. An ethical criterion is provided for assessing when responsibility exceeds risk: "The benefit the patient will gain outweighs any harm the nurse might incur and does not present more than minimal risk to the health care provider." An assessment of benefit to the patient from receiving CPR can only be made on a case-by-case basis. Does the same hold true for risk to the nurse in administering mouth-to-mouth resuscitation? If there were a widely agreed-upon threshold for minimal risk, an objective answer could be given to this question.

Minimal risk has been defined in one context, that of biomedical research involving human subjects: "The probability and magnitude of harm that is normally encountered in the daily lives of healthy individuals, or in the routine medical, dental or psychological examination of health individuals."[4] There is no doubt that the *magnitude* or harm to someone who becomes infected with the AIDS virus is excessive and unreasonable by any ethical standard. However, the problem in seeking to determine whether performing CPR poses more than minimal risk is that the *probability* of acquiring HIV infection remains unknown.

A third position paper issued jointly by the Health and Public Policy Committee of the American College of Physicians and the Infectious Diseases Society of America contains the following statement: "The American College of Physicians and the Infectious Diseases Society of America urge all physicians, surgeons, nurses, other medical professionals, and hospitals to provide competent and humane care to all patients, including patients critically ill with AIDS and AIDS-related conditions. Denying appropriate care to sick and dying patients for any reason is unethical."[5]

In its rationale for this position, the article cites then-current statistical data about transmission, Public Health Service guidelines as of 1985, and the fact that physicians and nurses are charged by the ethics of their healing profession to treat patients with all forms of sickness and disease.[6] There have been no significant changes since 1986 in the statistical rate or modes of acquisition of HIV infection by health care workers and no major changes in Public Health Service guidelines for the protection of health care workers. It can be inferred, then, that this position paper obliges physicians and nurses to perform mouth-to-mouth resuscitation on all patients for whom that treatment is appropriate.

This conclusion is further supported by reference to the data about transmission through casual contact. Ongoing studies of people exposed to HIV-infected individuals in the same household—for example, sharing a bathroom and eating utensils—continue to show no evidence of transmission.[7] Yet it is doubtful whether mouth-to-mouth resuscitation should be subsumed under the heading of casual contact. The incidence of blood in the saliva of patients in emergency situations, and the prolonged contact between the mucous membranes of the resuscitator and the patient's saliva, make CPR significantly different from the usual casual contact found in households.

Two different approaches can be taken to the problem of the risk of performing mouth-to-mouth resuscitation. According to one viewpoint, the risk to the resuscitator is the only factor that need be considered in determining whether an obligation exists to perform emergency CPR when protective mouth equipment is not immediately available. The highest risk is faced by a resuscitator with mouth sores about to perform mouth-to-mouth resuscitation on a patient known to be HIV infected and having bloody saliva. An equivalent or greater risk is presented by the neonate described in the opening case scenario. A hierarchy of descending risks could be devised, using the known facts and probabilities of transmission. A moral choice would still have to be made to decide where to draw the line between unreasonable and acceptable risk.

Another approach requires balancing the likely risks to the resuscitator against the potential benefit to the patient from receiving emergency resuscitation. This is the approach taken by the ANA Committee on Ethics, whose statements are formulated in terms of risk versus responsibility. It is perfectly evident that health professionals are not obligated to provide valueless care. Where to draw the line between marginally beneficial treatment that does not call for caregivers to place themselves at risk and benefit to a patient that does require some balancing against risks to the resuscitator is a line-drawing problem analogous to that of demarcating acceptable from unreasonable risk.

Ezekiel Emanuel, a physician-ethicist, identifies four considerations having potential ethical relevance to physicians' obligation *in general* to treat patients with AIDS.[8] In addition to risk to the physician and benefit to the patient, Emanuel mentions obligations to other patients and obligations to self and family as factors cited by physicians in refusing to treat patients with AIDS. He concludes that the only condition justifiably limiting physicians' obligation to care for patients with AIDS is pregnancy. Because contracting AIDS also places the unborn child at a high risk—up to a 50 percent chance—of being infected with HIV, Emanuel argues that the threshold of excessive risk should be lower for pregnant physicians. Even in that case, however, the conclusion may change as new data become available. Recent evidence suggests a lower rate—25 to 30 percent rate of infection in infants born to HIV-infected mothers. Is that threshold low enough to eliminate the exception of pregnant physicians from the general obligation to treat?

Since few, if any, ethical obligations are absolute, the need exists to delineate legitimate exceptions in a consistent manner. It would be useful to provide an objective standard for demarcating reasonable from excessive risk.

Emanuel is correct in noting that "the balancing cannot simply be a matter for the individual conscience."9 Yet the standard he proposes—"professional and social expectations"—is fraught with problems of its own.

Two problems surround the use of professional and social expectations as a guide to health care professionals' obligation to perform mouth-to-mouth resuscitation on HIV-infected patients. The first problem is how to ascertain accurately what those expectations are, given the wide variety of opinions voiced by professionals and the public. The second problem is whether people's expectations are the most appropriate basis for determining the scope and limits of professional obligations. Some surgeons have deemed the risk to them of operating on HIV-infected patients to be excessive, while other surgeons continue to operate on such patients. Some parents have refused to send their children to a school where children with AIDS are enrolled, while other parents accept the evidence that there is virtually no danger of transmission through casual contact and act accordingly. Which set of expectations should we use as a guide in cases of conflict? And how can expectations regarding the obligation to perform mouth-to-mouth resuscitation be ascertained? Professional and social expectations might not prove to be a consistent guide or one that is ethically reliable.

Emanuel uses another, more promising approach. Comparing the risks incurred by physicians in various specialty areas with those undertaken by workers in different occupations—for example, firefighters—he takes the bold step of drawing the line between a risk that is acceptably high and one that is excessive. The choice of an occupation, such as firefighter or driver of a truck with explosives or inflammable substances, carries with it a level of risk known in advance to be higher than that of other jobs. Taking risks that are acceptably high would seem to be implied by choosing medicine as a profession. But it is not obligatory for physicians to assume risks that cross that line and fall into the excessive-risk category.

Emanuel gives several examples. "Firefighters are not normally considered obligated to enter a blazing house to rescue the inhabitants if such an attempt would almost certainly result in serious bodily injury. . . . Similarly, the [American] military does not send soldiers on suicide missions, and for high-risk operations it requests volunteers."10 A risk that is high but not excessive is that faced by firefighters in the normal course of their work: "Recently, firefighters on the front line in Boston have faced a risk of death of about 0.5 percent during each of the worst years, and of 0.2 percent in normal years."11

Other examples of high risk in a health care setting have been suggested.12 "An example would be the rescuing of an injured client from a dangerous situation such as a burning room. In other circumstances, only other persons specifically trained should intervene, such as calling upon police personnel to deal with a client in possession of a lethal weapon."[13] This latter illustration does not help much in the case of CPR, since it is the specifically trained individual whose professional duty is under scrutiny.

Emanuel offers an example of excessive risks to a physician treating HIV-infected patients. Based on data provided by one orthopedic surgeon at San Francisco General Hospital, he calculates her risk as 12 percent per year of contracting HIV infection from a work-related puncture, with a five-year cumulative risk of 49 percent.[14] An example of a risk that is high and "borders on the excessive," according to Emanuel, is that of emergency department surgeons reported in one study with a 2 percent annual risk of contracting an HIV infection from performing major surgical operations. This is comparable to the risk faced by soldiers at the height of the war in Vietnam in 1968: an annual risk of less than 3 percent of dying in combat. The intuitive response to these figures is that surgeons and other health professionals should not be obligated to undergo risks equivalent to those of soldiers in combat.

This analysis suggests that there are limits to the moral obligations of health care workers to deliver care to HIV-infected patients. The outer limits are those in which the risk to the workers is *excessive*. Although such cases are rare, physicians have the moral option to refrain from acting in ways that present risks of harm that would be unreasonable for anyone to undergo. What magnitude of risk is excessive and how to interpret "reasonable" are, unfortunately, questions to which only subjective answers are forthcoming.

A different distinction is sometimes made in discussions about physicians' obligation to treat HIV-infected patients. Doctors inquire whether they are ethically required to perform elective procedures that may place them at risk of HIV infection. However, the term "elective" is ambiguous. Sometimes it is used to refer to any treatment that is not an emergency or, at least, not urgently necessary for the patient's life or health. At other times, it is used to refer to surgical treatments that are merely desired but not needed for life or health. Examples in this category typically include cosmetic surgery, such as nose jobs, face lifts, and the like, or a procedure like liposuction, for which nonsurgical alternatives (diet and exercise) exist.

Surgeons at one ethics conference presented the case of a homosexual man with HIV infection who was seeking treatment for impotence. One treatment they routinely perform on patients with this problem is a penile implant. Urologists at the conference questioned whether they had an obligation to do the procedure on this patient. They defined the treatment as elective and cited two very different reasons for why they were not obligated to offer the treatment. First, they cited the risk to themselves of acquiring HIV infection; second, they said they should not be "complicit" in the spread of AIDS. "Fix this gay man's inability to have sex, and he will be furnished with a lethal weapon. He'll be equipped to go out and spread the disease to others, and we as physicians have an obligation to the uninfected population not to do anything that would increase their risk."

The first reason offered by these urologists falls into the same category used to justify withholding treatment from HIV-positive patients in general: the risk to the surgeon. It is complicated, however, by the urologists' contention that the surgery is elective, and therefore, that they are not ethically obligated in the same way as if the procedure were a lifesaving one like resuscitation. This line of reasoning recalls the ANA's position that "accepting personal risk which exceeds the limits of duty is not morally obligatory; it is a moral option." The ANA's position on the nurse's obligation is framed in terms of a balance between benefit to the patient and risk to the provider: "The benefit the patient will gain outweighs any harm the nurse might incur and does not present more than minimal risk to the health care provider."

In discussing this position earlier, I said that such assessments have to be made on a case-by-case basis. That forces us to evaluate the benefit to the impotent patient of restoring his sex life and to weigh that benefit against the risk to the urologist of performing the penile implant. I asked the urologists at the conference whether the procedure typically involved a lot of blood, and they replied that it did. I then asked about the risk of punctures through latex gloves in the course of doing the procedure, and they replied that this risk always exists during surgery. Based on these two responses, the next task is to determine whether the risk to the surgeons is more than minimal. If it is, the problem is resolved using the ANA formula: Weighing benefits against risks has to be done only if the procedure *does not* present more than minimal risk to the health care provider. If the risk is not more than minimal, then the risk-benefit problem must be solved.

What is the criterion for determining when a risk is minimal? As noted earlier, there is, unfortunately, no objective criterion in the case of health care workers. The risk to urologists of becoming HIV infected in the course of doing penile implants does not appear to be greater than the risk faced by most other surgeons doing different operations and is probably much lower than that faced by orthopedic surgeons. In that case, it would appear necessary to balance risks against benefits to patients undergoing elective procedures such as penile implants. There is simply no objective basis on which that balancing could be done. When benefit to one person has to be weighed against risk to another person, an inevitable subjectivity enters the picture. The one who is subject to the benefit will undoubtedly weigh it differently from the one who is subject to the risk. When judgments must rely on irreducibly subjective elements, there is no objectively right or wrong answer.

This variation in personal values emerged in two conversations I had following the conference with the urologists. One colleague who was not at the conference expressed sympathy for the position taken by the urologists. This physician normally argues that surgeons do have an obligation to operate on HIV-infected patients, yet in this case he questioned whether the benefit to the patient justified the risk the urologists would have to take. "After all," the physician contended, "impotence is not a life-threatening condition, and the patient's overall health is not likely to deteriorate if this condition isn't corrected."

A second colleague, a psychologist who works in the department of urology, held a different view of the benefits. The psychologist's professional work has led her to see the prominent role sexuality often plays in people's lives. She did not evaluate the magnitude of the benefit in terms of risks to the patient's life or overall health, but rather in terms of the quality of life as determined by the patient suffering from impotence. Whereas the physician tended to view the matter according to strictly medical criteria, the psychologist included the patient's capacity for sexual satisfaction and emotional well-being in her assessment of benefits. The psychologist also made the interesting observation that urologists must, in general, hold sexual functioning to be an important aspect of life, since much of their practice is devoted to helping men who have sexual dysfunction.

The second reason given by the urologists for why they are not obligated to do the penile implant—that they would be complicit in the spread of AIDS—invokes a perfectly sound ethical principle but misapplies the principle to this case. The ethical principle is a variation of the harm principle: An action that would be obligatory under normal circumstances is not obligatory, or may even be prohibited, in cases where performing it is likely to lead to harm to others. The urologists were contending that even if they are normally obligated to do penile implants when an impotent man seeks treatment (including an impotent gay man), the obligation ceases when the patient is HIV positive, since once treated, he will contribute to the spread of AIDS and thereby cause harm to others. In this case, the threat of harm is not to physicians but to the public.

The urologists who presented this argument (it was not endorsed by the entire group) made at least two factual assumptions, which could very well have been false. These were (1) that the patient was not in a monogamous relationship with a partner who knew or would be told by the patient that he was HIV positive and (2) that if the patient was not in a monogamous relationship, he would behave recklessly and withhold information or otherwise endanger future sex partners. Since proper application of the harm principle requires a realistic assessment of the probability that the envisaged harm will actually occur, a conclusion regarding the urologists' obligation cannot be reached in the absence of the relevant factual information.

But even with that information, a question still remains about the nature and extent of physicians' obligation to society or to unknown, unidentified individuals. Reasoning analogous to that of the urologists might lead to the conclusion that a physician does not have an obligation to treat diseases or injuries in violent criminals or sex offenders, since by curing these individuals the physician is enabling them to go out and commit more dastardly deeds. The response to this line of reasoning is that crime prevention is not the province of physicians. But what about disease prevention? Isn't that the province of physicians?

A public health role is indeed an appropriate one for physicians to assume. But when that role comes into direct conflict with a doctor's ordinary obligation to provide care or treatment to an individual patient, a careful assessment is required before leaping to the conclusion that an obligation to the public outweighs an obligation to the individual patient. The urologists who argued that they would be complicit in the spread of AIDS if they helped the impotent gay patient overcome his sexual dysfunction had no warrant for their belief. These physicians were more than likely responding to a stereotype of gay men or expressing hostile or prejudicial attitudes against gays. They were not justified in refusing to treat the man on these prejudicial grounds.

A new twist in the ongoing controversy over physicians' obligation to treat HIV-positive patients has arisen, perhaps ironically, out of growing calls by the public for testing of doctors, dentists, and other health care workers. Following the discovery that a dentist in Florida who had AIDS was responsible for infecting five of his patients, demands that physicians, dentists, and other health care workers be tested for HIV infection became widespread. Health professionals are now being perceived as potential enemies of patients. It is hardly surprising that if patients come to view their physicians as potential enemies capable of infecting them with a deadly disease, physicians will retaliate and escalate their request that patients be tested for HIV infection.

The demand of patients that physicians be tested poses a new threat to health professionals: a threat to their very livelihood. The Centers for Disease Control has refrained from mandating HIV tests for physicians but has urged that doctors be tested on a voluntary basis. Those who are HIV positive should then refrain from performing invasive procedures.

If the risk to physicians of acquiring AIDS from patients in ordinary medical practice is minimal, the risk to patients from HIV-infected physicians is even less. In almost all areas of medical practice, the contact between doctors and patients falls into the category of casual contact. This is less so in surgery, yet what would have to occur to cause the risk of a patient's acquiring the infection from a surgeon is direct bleeding into a patient's open cavity. Of course, this is possible and probably does happen on occasion. The public's response to the prospect of being infected by a doctor or dentist is not, however, a rational one. It is, like most of the reactions to AIDS-related phenomena, highly emotional and based on poignant tragedies like that of Kimberly Borgalis, the attractive young woman in Florida who was a patient of the dentist who was alleged to have infected her and four other patients, and who subsequently died of AIDS.

The backlash on the part of physicians is not surprising. Dr. Richard Howard, a professor of surgery, remarked: "If I have to be tested and no hospital will allow me to practice if I'm positive, then I'm going to start thinking, 'How can I reduce my risk?' The only thing I do that's risky is operate on patients who are HIV positive. And so I think some surgeons may decide to stop."[15]

A similar response came from Dr. William Schecter, a trauma surgeon at San Francisco General Hospital, who has encouraged other physicians to take care of patients with AIDS. Dr. Schecter reacted to the passage of a bill in the U.S. Senate, sponsored by Senator Jesse Helms, that would have forced health-care workers performing invasive procedures to inform patients if they had the AIDS virus or face prison terms and fines. Schecter was quoted as saying: "I feel really betrayed: Here I am on the front lines treating everyone and now Jesse Helms says if I acquire HIV infection I'm not going to be able to support my family."[16] Dr. Schecter said that his colleagues are now asking whether they should continue to take care of AIDS patients. In addition to the threat of physicians' exposure to the virus, which has existed since the beginning of the AIDS epidemic, there is now the added threat to their livelihood. The Helms legislation was rejected eventually by a House-Senate conference committee appointed to negotiate the measure.

As in all situations in which fear drives human behavior, to encourage people to respond on the basis of a rational assessment will have little effect. While it is surely rational to fear an outcome like death, it is not rational to ignore the low probability of death from a particular cause. The risk of death from many ordinary activities in daily life is greater than the risk of becoming infected and dying from an encounter with an HIV-infected doctor or dentist. Yet most people do not choose their every action based on a calculated zero probability of the risk of death. Both sets of demands—physicians' request that they be allowed to test patients without the patients' consent and the public's call for testing physicians—are guaranteed to make adversaries out of doctors and patients. As long as patients and physicians view one another as posing a threat of harm, there is little hope for improvement in the physician–patient relationship.

The Abusive Patient

In addition to the risk of becoming infected with the AIDS virus, there are other kind of situations that pose a threat of harm to physicians. One category comprises threatening, abusive, or violent patients who intentionally try to harm health care workers or who are deliberately combative. Although such patients are uncommon, they do exist. Physically combative or abusive behavior can be a result of drug or alcohol intoxication, psychosis, sociopathy, or a disposition to behave violently. In a hospital, such patients can pose a threat not only to doctors and nurses, but possibly also to other patients. The following case is illustrative.

The patient was a 28-year-old male with a long history of intravenous drug use. He had had many hospitalizations, some for deep vein thrombosis, others for bacterial endocarditis, and still others for additional complications stemming from his drug use. He was admitted this time for deep vein thrombosis, complaining of pain in his lower extremities. He demanded Demerol for the pain, as he had on numerous previous occasions at this hospital. When he was refused the Demerol on the grounds that it was not indicated for his condition, he refused the Coumadin prescribed for his deep vein thrombosis. It was impossible for physicians to maintain therapeutic levels of Coumadin because of the patient's noncompliance. As a compromise, he was given 75 milligrams of Demerol instead of the 100 milligrams that he had been given on previous admissions and that he was now demanding.

The patient became violent and abusive. He again demanded his "full dose" of Demerol, which was refused him. He threatened to hit a medical student over the head with his cane. He threw a wheelchair around the ward, and hospital security was called. The outburst ended and the patient remained in the hospital, eventually accepting the Coumadin but refusing to remain in bed with his leg raised. The staff reported that this was not the first such episode, and said that rumor had it that the patient had been refused admission to other emergency rooms in the city because of his attitude and the fact that he demanded Demerol as a condition of treatment. In this instance, the patient threatened to sue the doctors and the hospital for withholding "proper" medical treatment.

Although noncompliance alone does not count as a sufficient reason to discharge a patient from the hospital, this case has two added features: First and foremost, the patient's violent and abusive behavior threatened health care workers and interfered with the care of other patients. Second, patients have a right to refuse a recommended treatment, but they do not have a right to demand inappropriate treatment such as medication that is not indicated for their condition. It was evident to the staff that this demanding and manipulative patient was less interested in receiving therapy for his medical condition than in obtaining what he could get away with by bargaining with the physicians.

Fortunately, this is not typical of the majority of cases in which patients seek to manipulate physicians or make unreasonable demands, but it does illustrate an extreme case that makes it possible to determine where physicians' obligations cease. The physician is not obligated to treat a patient under conditions dictated by a patient who is holding the medical team hostage to his or her demands. And a hospital is not morally obliged to keep an ambulatory patient who represents a danger both to the staff and to the proper care of other patients. Such cases may be rare, but they demonstrate that there are limits to the obligations of physicians to care for patients.

Another abusive patient, R.L., had a long history of multiple drug use. He was known to become more verbally abusive, more threatening, and more destructive after drug intake. R.L. was mildly mentally retarded, had spent his entire childhood in the care of institutions, and in recent years had been homeless. He voluntarily entered the psychiatric unit of a general hospital because he was homeless, depressed, and had tried to burn a relative's house. Because psychiatric patients can remain for only a limited time on the psychiatric ward of a general hospital, when his time had expired he was referred to a long-term psychiatric facility.

Once in the facility, R.L.'s behavior deteriorated further. He threatened other patients and staff with physical harm. He was found to possess an improvised sharpened instrument, with which he threatened a staff member, and refused to reveal how he had obtained it. Following that episode, R.L. got into a confrontation with the evening nurse on duty. After threatening to harm the nurse, he destroyed her office. He had a confrontation with a woman patient in the group therapy program and threatened to assault her off the hospital premises. He also threatened to assault one of the psychiatrists who was examining him for a physical condition, and the male nursing staff had to come to the psychiatrist's aid.

Most recently, R.L. was informed by a psychiatrist at the hospital, on behalf of the treatment team, that his request for off-ward privileges had been denied because of repeated threats he had made recently to several staff members. This psychiatrist contended that R.L. was not mentally ill, but that his threats to harm others were manipulative and coercive in nature and intent. Based on that assessment, the psychiatrist denied privileges to R.L.

R.L.'s response was to announce that he would "destroy the ward" and that the staff would see what he was capable of doing. He rejected the attempt by several staff members to talk with him and immediately thereafter began to destroy property on the ward. Within a matter of minutes, R.L. ripped two sets of emergency lights off the wall, broke the hydraulic door control on the front door of the ward, and disengaged a lighted exit sign from the ceiling. When his psychiatrist attempted to talk with him, R.L. challenged him to "try and stop me" and kicked the doctor's door. At that point, security officers were summoned to help seclude R.L. After spitting on the unit chief, the patient informed the safety officers that they should be prepared to come to the ward every day. He had every intention of continuing to act this way.

The psychiatrist in charge of R.L.'s care wrote the following note in his chart:

> This patient is currently non-suicidal and non-homicidal. He exhibits no symptoms of psychosis. It should be noted that the patient recently destroyed property on this ward, when he became very angry because he did not receive privileges. This destructive behavior is considered willful and goal directed. He is not considered mentally ill and does not require medication. Please note as well, the patient is capable of handling the stress of a booking by police, and I believe that this will not adversely affect his mental status.

This case posed a troubling dilemma for the psychiatrist who served as director of psychiatric services at the hospital. The director was torn between his obligation to serve the best interests of R.L., his duty to provide a safe and conflict-free institutional environment for other patients, and his obligation to protect his own staff from potential physical harm. The director prepared the following summary when R.L.'s case was presented at a conference:

> The patient lacks the social, educational and vocational skills to survive in the community. Additionally he is extremely manipulative and provocative, which makes others vulnerable. He refuses, however, to participate in programs that would help him to obtain the skills he would need to be discharged from the hospital. At the same time, he said that he wants to stay in the hospital for life. He is not psychotic, and does not meet the criteria for continued hospitalization. He also takes advantage of other, more vulnerable patients.
>
> Although this patient has no mental illness, he does have serious disabilities of a nature that would prevent his integration into the community. Should we discharge him, or should we continue to attempt to engage in rehabilitative treatment?
>
> Is it ethical to take the position that if the patient cooperates with the treatment that we provide, we will treat him as a patient; if he does not, we will treat him as a criminal?
>
> How do you balance the wish of this patient to remain in the hospital with the rights of other patients to a safe environment?
>
> Within this community, we are tolerant, supportive, provide social interactions, etc., to a degree that frequently is not available in the communities to which we discharge. Is it ethical to discharge a patient perceived as being impoverished with the possible consequence of an early readmission?

The case of R.L. is a genuine dilemma, as it poses two alternative courses of action, both of which are ethically problematic. Moreover, each alternative can be justified by appealing to a sound and generally accepted ethical principle. The harm principle can be used to support R.L.'s discharge, since his remaining in the hospital poses an ongoing threat of physical harm to other patients and to staff. The same principle can also justify not discharging R.L. to the community, since when he was given a home leave he had a confrontation with one of his friends, surrendering a steak knife on his return to the hospital. He also got into a confrontation with his girlfriend while off the ward; he slapped her and threatened to kill her. So, the same principle can be used to justify either of the alternative actions.

As for the principle mandating that physicians must act in the best interest of the patient, it is unclear just which course of action serves that end. R.L. claims he wants to remain in the hospital, yet he is unwilling to engage in institutional programs, which offer the only prospect for benefiting him. Moreover, his behavior on the ward has resulted in his having to be placed in seclusion for the protection of others as well as hospital property. Being forced to remain in seclusion rarely helps a person who doesn't want to be there. The other alternative, discharging R.L. from the hospital, has the consequence of throwing him into a world that he is ill-equipped to deal with and will likely result in his being readmitted to the psychiatric hospital or landing in jail.

The case of R.L. has no resolution that is clearly ethically right. Defenders of individual liberty will contend that it is wrong to detain people in coercive environments against their will, yet it was not clearly against R.L.'s will to be in the hospital. Paternalists might claim that the patient is better off in a structured environment, yet he refused to participate in rehabilitative programs and his destructive behavior landed him in seclusion. Libertarians or retributionists would argue that R.L. is not mentally ill and therefore should not be in a hospital at all, willingly or unwillingly. He should be discharged, and if he engages in criminal or antisocial behavior, he should be handled by the criminal justice system.

An additional consideration was brought out at the case conference. R.L. had been in institutions since he was a very small child. He had not been brought up in a family setting and was thus a product of custodial care. The system made him what he is today. He is what he is because of having been raised in medical institutions. Instead of abandoning him now that he is an uncontrollable adult, doesn't that same system owe him something in return? Even if the answer to that question is "yes," it remains unclear just what is owed to R.L. and what is the proper way of discharging that obligation to him.

The director of psychiatric services had to weigh all these factors and decide. The director was thoughtful, reflective, and eager to do the right thing. And he was young. In the end, he decided that his greater responsibility, as a psychiatric administrator, was to protect the other patients in the hospital, as well as his own staff. Since the benefits to R.L. of remaining in the hospital did not clearly outweigh the benefits of being discharged, that would be the least bad solution. In a situation with so many competing factors, the director's decision was as ethically acceptable as any.

After he announced his decision at the end of the conference, this young psychiatric administrator was praised by a senior colleague who had been his teacher and mentor. The senior psychiatrist praised him not only for his sound ethical reasoning, but also for not allowing another consideration to dominate. R.L.'s discharge from the hospital would most likely lead to his getting into trouble again, and possibly to adverse publicity for the psychiatric hospital and the director himself. By discharging a patient likely to behave antisocially and perhaps even criminally, the director was exposing himself to criticism or worse from his superiors in the state's department of mental health. His action took courage, the courage of acting out of his convictions rather than out of fear of censure from state bureaucrats. For that courage, as well as for his thoughtful analysis, he deserved the praise he received from his more experienced colleague.

Patients Who Are Dangerous to Themselves or Others

Although R.L. did not have a mental disorder that fit a precise psychiatric label, his case is similar in many ways to that of psychiatric patients who pose a threat. He surely lacked the ability to control his impulses. Debates concerning the rights of psychiatric patients have long raged within the psychiatric profession as well as outside it. Some mentally ill patients are judged to be in need of care, custody, or treatment; may such patients be confined against their will? This is the classic debate between paternalists, who argue that coercion is ethically permissible in order to prevent people from harming themselves, and civil libertarians, who argue that it is wrong to interfere with people's liberty unless they have committed a crime or have a high likelihood of harming others.

Controversy has also surrounded the proper treatment of people judged dangerous to others. One side argues that although taking away people's freedom for their own good cannot be justified, interfering with their liberty to protect others from harm is justifiable. That ethical justification is based on the "harm principle" and differs from a paternalistic justification for coercion. The opposite side, consisting of strict libertarians, contends that the only time it is permissible to interfere with the liberty of people in a free society is in response to criminal acts they have already committed. "Preventive detention" is not an acceptable practice, according to this position, and psychiatrists who are allies of the state in getting people locked up have become agents of social control rather than the healers their training and profession equip them to be.

These issues are old ones. No consensus has been reached either within the mental health profession or among the general public. I am not going to rehearse the arguments and evidence surrounding the opposing positions in these debates, but will instead explore another justification psychiatric administrators frequently use for constraining the liberty of patients. That justification has little to do with the rights of patients, their best interests, or the safety of innocent members of society. Rather, it pertains to the psychiatrists' fear of making a mistake and the consequences they envision to themselves. Once again, the motive that drives behavior is self-interest rather than patient-centered.

At a monthly conference for psychiatric residents and postgraduate fellows, current or recent cases are presented by the trainees, followed by open discussion. In the midst of a discussion of the central ethical and legal issues, a senior psychiatrist often refers to a reason for action as an "administrative concern." That has become a clue that what ultimately drives decision making in a psychiatric facility are the interests of the hospital and its administrative personnel.

One such case involved Mr. Norton, a man in his early seventies, divorced for 30 years, who was brought to the hospital's emergency room following an injury to his leg. The emergency room surgeon who sutured Mr. Norton's leg called for a psychiatric consultant because the patient reportedly stated that he had injured himself in a suicide attempt. In the psychiatric interview that followed, Mr. Norton denied having made a suicide attempt and claimed that the injury was accidental. The psychiatrist found him to be "uncooperative with further interview," during which the patient became angry and called the psychiatrist obscene names. The next day, Mr. Norton was admitted to the psychiatric inpatient ward.

Once on the inpatient ward, Mr. Norton was examined by more than one psychiatrist. The doctors notified the patient's son, with whom he had only occasional contact. The son provided the additional information that his father had a long history of drinking a quart or more of whisky per day. Mr. Norton's landlady had tried to evict him some years ago because of his large collection of cats that had become unbearable for his neighbors. The son also mentioned that his father had been admitted to another hospital a year before because of apparent suicidal ideations. The patient gave a history of trying to hang himself as a youngster and of having changed his mind; he reported another suicide attempt 25 years ago but, again, a change of mind. His physical examination in the hospital revealed a mild vitamin deficiency and impairments of hearing and vision.

The psychiatrist's assessment was that Mr. Norton was of average intelligence, had poor memory and insight, was unable to think abstractly, was uncooperative with the psychiatrist who performed the interview, and had a number of "morbid false beliefs" and some delusions. He was not found to be either suicidal or homicidal. However, based on the psychiatrist's concern that he might come to harm, Mr. Norton was detained in the hospital for 14 days, the maximum permitted by law following an emergency admission. At the end of this period, the patient must be evaluated and a decision made either to release him or to commit him in a procedure known as "2-PC," requiring a two-physician certificate that can keep a patient hospitalized for 45 days, at which time another determination must be made.

After the initial 14-day period, Mr Norton was "2-PC'd." At the time of the mandatory 30-day review, his mental status exam remained about the same as before. Psychiatrists decided to increase the medication that he was being given and to add another drug if his symptoms continued. It would soon be time to decide whether to seek his admission to a long-term psychiatric institution, and the psychiatric resident was in conflict about the proper course of action. He posed two chief questions at the conference: Should Mr. Norton have been hospitalized under emergency status in the first place? And is continued involuntary hospitalization really necessary?

In the discussion that followed, everyone agreed that the initial emergency hospitalization was justified. One psychiatrist acknowledged, however, that emergency room doctors frequently call psychiatrists in whenever there is any doubt about a patient's mental condition or the possibility of suicidal behavior. This practice admittedly stems from a "cover your behind" mentality on the part of emergency room personnel, but was thought justified anyway because emergency room physicians are generally surgeons and trauma specialists, not especially trained in techniques of psychiatric interviewing. Everyone agreed that until details about the patient's psychiatric history and condition could be clarified, an emergency admission to the hospital was warranted.

Disagreement began with discussion of the second question: Is Mr. Norton's continued involuntary hospitalization really necessary? Stressing the fact that ethical issues are not decided by vote, I asked for a show of hands. The majority of the young psychiatrists unhesitatingly opted for the paternalistic solution. One commented that with a psychiatric history of suicide attempts, they had to act on the safe side. Another commented that the patient's vitamin deficiency showed him to be in need of care for the sake of his health, and as physicians they were obligated to see that he received proper nutrition. Still another noted that if he were discharged, he would not comply with his psychiatric medications; for that reason, it was justifiable to keep him hospitalized.

Three participants voted on the opposite side: that continued hospitalization was not right. The psychiatric resident who presented the case at the conference reluctantly voted on this side but sought assurance from his senior colleagues. The two experienced psychiatrists claimed that there was no psychiatric or ethical justification for detaining the patient. His history of suicide attempts was fuzzy at best, and in any case there was no real evidence of suicidal behavior at the time of this hospitalization. The fact that he had a vitamin deficiency did not distinguish him from the vast number of poor or elderly people, and although his nutritional status was unfortunate and ideally should have been improved, it could not serve as a reason to force him to remain in a psychiatric hospital. As for the antipsychotic medication administered to Mr. Norton, it was not clearly efficacious for his condition. The decision to increase the dose and add a different drug represented the treating psychiatrist's desperate attempt to do something in the face of failure of the present medication and dosage level.

When I asked the assembled group to vote, I phrased the question as it had originally been posed by the psychiatrist who presented the case: Is Mr. Norton's continued involuntary hospitalization really necessary? The psychiatrist/attorney replied: "Are you asking what's right or what I'd do?" Having made that distinction, she went on to say that she believed it was wrong to seek continued hospitalization for this patient but admitted that it was what she would do anyway. With some embarrassment, the senior psychiatrist at the seminar said he would do the same.

The reason, they both acknowledged, came back to these administrative concerns. Suppose that some harm did come to Mr. Norton—a return of his suicidal ideation, a worsening of his nutritional status, or eviction by his landlady because of all those cats. It wouldn't look right for the hospital and its psychiatric staff if they had discharged a patient who turned out to need care, custody, or treatment. There could be bad publicity, the patient's son might initiate a lawsuit, or the hospital or the physicians could be cited by the state's department of mental health. Who knew what bad things might happen? The psychiatrists who used these administrative worries as a reason for detaining the patient were not invoking paternalistic reasons. They acknowledged that paternalism was not ethically justified in the case of Mr. Norton. The harm principle provided no justification, since there was never any question of Mr. Norton's being a danger to others. The threat posed by this patient was seen by the psychiatrists as a threat to themselves and to the hospital, albeit a vague and remote threat.

Throughout the years this seminar has been conducted, cases have been presented in which the reasons psychiatric patients are admitted or discharged have only partly to do with their psychiatric illness. In many cases it is the interpretation of the patient's family or that of the police that governs whether the patient will be admitted initially. For example, when the police bring a patient in handcuffs to the psychiatric emergency room, the patient is always admitted. If the family is unwilling to accept the patient back into the home, that becomes a reason for detaining the patient in the hospital even if he or she cannot benefit from further treatment as an inpatient. Most of the time, decisions are made on the basis of legal implications: A provision of the New York State mental health law requires psychiatrists to admit only those patients who pose a danger to themselves or others and who can benefit from psychiatric treatment.

Psychiatrists who serve the dual functions of clinician and administrator are plagued by the perennial problem of "double agency." As a clinician who makes evaluations, and who engages in diagnostic and therapeutic decisions, the psychiatrist's primary obligation is to the individual patient. When a psychiatrist who performs those functions also administers the unit or floor on which patients are housed, genuine obligations to others besides the patient emerge. In cases where patients are evaluated as dangerous to others, a legitimate obligation to society (preventing avoidable harm) can conflict with an obligation to the patient (maintaining respect for autonomy).

Conflicting obligations of this sort are not new for physicians, nor are they unique to psychiatry. However, when they incorporate administrative concerns into their judgments about detaining or discharging patients, psychiatrists begin to act as risk managers as much as they do as clinicians responsible to their patients. One psychiatrist stated his views about the relationship between good ethical practice and risk management. The ethical standards regarding the care and treatment of patients may be kept high, with the understanding that ethical ideals can only be approximated; or they may be set lower to conform more to the demands of reality. Keeping the standards high comports with good ethics, but making them lower is more desirable from a risk management perspective. This is because lower standards are easier to comply with and, therefore, reduce potential liability.

The two psychiatrists at the conference who perceived a difference between what they thought was right and what they admitted they would do in the case of Mr. Norton were attentive to ethical issues and at the same time attuned to practical realities. Those realities are even more compelling in psychiatry than in general medicine because of the possibility that a psychiatric patient is homicidal, a rapist, or a child molester and thus poses a genuine danger to others. Added to these unsavory consequences are the adverse publicity and damage to the reputation of those particular psychiatrists and perhaps to the profession as a whole.

To their credit, the psychiatrists who made the distinction between what was right and what they would do did not lump all of these concerns under the heading of ethics. They were careful not to confuse administrative concerns with those that properly comprise ethics—acting in accordance with principles. The decision they finally made had little to do with ethics and much to do with how they viewed their own professional interests. A patient who becomes a threat to those interests is seen as an adversary of physicians, with the result that doctors act to protect themselves rather than to serve their patients' best interests.

Notes

1. Council on Ethical and Judicial Affairs of the American Medical Association, Report A, "Ethical Issues Involved in the Growing AIDS Crisis," (unpublished report) 1987, p. 4.

2. American Nurses Association, Committee on Ethics, "Statement Regarding Risk V. Responsibility in Providing Nursing Care," 1987, p. 1.

3. Ibid., p. 2.

4. 44 FR 47695, as cited in 45 CFR Part 46 (Washington, D.C.: Department of Health and Human Services, January 26, 1981), p. 8372.

5. *Annals of Internal Medicine*, vol. 104 (1986), p. 576.

6. Ibid.

7. Gerald Friedland, B.R. Saltzman, and M.F. Rogers, "Lack of Transmission of HTLV-III/LAV Infection to Household Contacts of Patients with AIDS or AIDS-Related Complex with Oral Candidiasis," *New England Journal of Medicine*, vol. 314 (1986), pp. 344–49.

8. Ezekiel J. Emanuel, "Do Physicians Have an Obligation to Treat Patients with AIDS?" *New England Journal of Medicine*, vol. 318 (1988), pp. 1686–90.

9. Ibid., p. 1690.

10. Ibid., p. 1687.

11. Ibid., p. 1688.

12. Sharon Jeanne Smith and Anne J. Davis, "How Much Risk Is Duty for the Health Care Practitioner? *Medicine and Law* vol. 5 (1986), pp. 29–34.

13. Ibid., p. 33.

14. Emanuel, "Obligation to Treat," p. 1688.

15. Elisabeth Rosenthal, "Angry Doctors Condemn Plans to Test Them for AIDS," *New York Times* (August 20, 1991), p. C1.

16. Ibid., p. C5.

7

Physicians as Fiscal Gatekeepers: Rationing at the Bedside

How would you feel if you knew that your doctor was holding back on your medical treatment in order to cut costs in the overall expenditures for health care? If your doctor decided that because the standard technique for early detection of cancer has been good enough for a decade you can do without an expensive new diagnostic procedure, would you be willing to accept the physician's decision to take the less expensive option without consulting you?

At present, we are constantly faced with reports of politicians, bureaucrats, and economists bemoaning the rising costs of health care in the United States. As citizens, most of us have become convinced that something ought to be done to control the inflationary spiral of health care costs. Don't doctors have a responsibility to society to help reduce spending for medical services? Our desire to get the best possible medical care when we are sick and, simultaneously, our feeling that less money should be spent for health care produces a tension that is not easily resolved.

When physicians come to believe that they are part of the problem, they tend to believe that they ought to be part of the solution. This tendency is exemplified in statements made by medical students with increasing frequency, such as: "This patient has been in the hospital for six months and on a respirator for four. Hasn't she cost society enough money already?" In teaching conferences that focus on clinical ethics, students and trainees now routinely mention the "costs to society" under the heading of ethical reasons for a particular therapy to be withheld or withdrawn from a patient. Many people claim that cost consciousness on the part of young physicians is a positive trend.

This recent trend suggests that physicians today feel an obligation to engage in rationing at the bedside. I have no hard data. And I am not contending that the vignettes I'm about to present are the equivalent of data or that a careful study would reveal that they are representative of a widespread practice. I have little doubt that there is increasing pressure on physicians to cut costs by rationing at the bedside and that it is already occurring. But whether deliberate rationing by doctors is a good move or a bad omen deserves serious contemplation.

What form does rationing by physicians take? One sort of rationing is based on the individual doctor's belief that a treatment cannot be justified on *economic grounds,* rather than on medical grounds, for a particular patient. An example described earlier was that of Lori, who was hospitalized for orthopedic surgery on her legs and feet to repair a birth defect. After surgery, the attending physician in the hospital refused to fill out the papers authorizing a home care worker on the grounds that "It is an abuse of Medicaid," and his opposition to having "our tax dollars pay for this."[1]

That example speaks for itself. To withhold proper care for a patient because she has borderline intelligence or because she has not been compliant with medical recommendations in the past is clearly a violation of medical ethics. But consider the following two cases, presented at a conference for third-year medical students doing their required clerkship in the department of medicine on the wards. Leading the conference along with me was a professor of medicine in charge of the student clerkship with whom I've been co-teaching this conference for 10 years.

This was the fourth ethics conference with this group of medical students. The student presenting the case reminded the group that a case discussed in an earlier session raised the question of when physicians may refuse to treat a patient (that case involved an HIV-positive intravenous drug user in need of a heart valve replacement). The student said he wished to address the question of physicians' refusal to treat once again, but for a different type of patient.

He described the ethical problem as one of conflicting obligations: physicians' duty to provide medical treatment versus their obligation to society. He characterized the two patients as coming from the "margins" of society. Both had been hospitalized for a long time at "great expense to the public," the student said. What are the limits to continuing to take care of these patients, especially given the poor financial status of the city, the state, and the nation?

In the first case, the patient was a 36-year-old man with a history of alcohol abuse and intravenous drug use. He was diagnosed as being in kidney failure as a consequence of his use of heroin and possibly also HIV disease, and had a large number of other medical problems. A full diagnostic workup was done, and the patient was found to have multiple infections, liver problems, and other diseases requiring treatment. While all this was being done, the patient had a sudden change in mental status and became unable to communicate. Renal dialysis was performed for his kidney failure, and he was transferred to the medical intensive care unit, where he was treated for pneumonia. Over the ensuing days and weeks he was given many treatments, and finally his mental status returned to normal. At the time of the ethics conference, the patient was mentally alert, without any fever, but "still HIV positive," the student said, and required regular dialysis for his kidney failure. He had been in the hospital for two and one-half months, taking up dialysis space. Although he now appeared to be ready for discharge, my physician colleague pointed out that the patient would have a hard time finding a dialysis unit that would treat him, since they don't like to accept HIV-positive patients for regular, ongoing dialysis.

The second patient was a 57-year-old woman who had been severely mentally retarded since birth. She was a permanent resident of a facility for developmentally disabled people, from which she had been transferred to this hospitalization. The medical student described the woman as having a very small vocabulary; she was unable to feed herself but assisted the feeder. She was normally in a wheelchair, but in the hospital she just lay in bed, moving little, if at all. The patient did recognize familiar faces and sang.

The woman had a long medical history, with many hospitalizations for infections, hypothermia, and anemia. Before being hospitalized this time she had refused to eat. She had many tests and evaluations, and after being treated aggressively for infections and anemia, was back to her baseline mental status, singing once again. She was discharged after being in the hospital for one month, to be followed on a regular basis in the clinic.

The student who presented the cases asked several questions tentatively, but with the conviction that they are ethical issues physicians today need to resolve: Does society owe everybody medical care? What about patients like these, who seem to test the limits by requiring prolonged hospitalization and many procedures? Who is to decide these questions? And how should physicians responsible for caring for such patients in the hospital behave?

The discussion that followed these two case presentations raised a host of familiar issues relating to the limits of medical treatment. But those issues are typically brought up in the context of patients who cannot benefit from continued treatment because they are comatose or in a persistent vegetative state (PVS) or who are terminally ill or in the process of dying and will not survive to leave the hospital. Debates about the obligation to begin or continue treating patients for whom treatment would be futile are increasingly common. But the two patients discussed at this conference did not fall into any of those categories. When the medical student characterized the patients as being from the "margins" of society, it was clear that he was referring to their perceived worth to society.

One student pointed out that neither patient worked, so neither was a contributor to society. Does society have an obligation to pay for the medical care of people who are a drain on its resources? It was easy to discern the underlying political convictions of the students who took up sides in this debate. The physician who co-teaches the conference noted that there was a distinction between the two patients and asked whether that mattered: The

first patient was perceived as not being a valuable member of society because
he was an alcoholic and drug user; the second patient might be said to have a
"low quality of life" stemming from her mental retardation. Another difference was that the first patient had AIDS, and therefore his overall prognosis for survival was very poor, whereas if the second patient received decent
medical care, she might live for another 20 years.

The majority of the students focused less on the different characteristics of
the two patients and more on the fact that their hospital care cost society so
much money. Someone noted that it costs $500 a day for a hospital bed on a
regular floor and $1000 in the intensive care unit. But the professor of
medicine replied that the cost of tests and medications administered to these
patients was far less than an array of different costly medical procedures given
to other patients. It was not the financial expense, taken by itself, that
motivated this discussion. It was the expense "wasted" on patients from the
"margins" of society, patients whose social worth might not entitle them to
drain society's precious resources in this way.

To be sure, not every student in the room endorsed the views of those
doctors-to-be who asserted that it is the physician's obligation to save society
money. Some were genuinely distressed at their colleagues' judgmental attitude, an attitude that in the future could readily determine which of society's disenfranchised members could live and which should be allowed to die
from lack of medical treatment.

Others argued that health care dollars would be better spent on prenatal
care, for keeping the hospital pharmacy open, for neighborhood health
clinics, and for other worthy causes. But when asked "Where do you think
the money would actually go, if not to care for these particular patients?"
most could not answer. One student accurately guessed: probably to treat
other patients like this one at this hospital. As a group, the students were
woefully ignorant about the details of health care financing: who pays for
what; which expenses are supported by taxpayers and which ones by an
insurance pool; and how the allocation of funds from federal, state, and local
governments is carried out. Despite this ignorance, a sizable number thought
it their obligation to act as fiscal gatekeepers, not only to save money that
might be wasted on futile treatments but also to save funds being spent on
patients of questionable social worth.

One group of students at this conference was embarrassed at the arrogance
and elitism of their peers. The other faction was genuinely angry—at the first
group, and also at my questioning the authority of physicians to decide, on
distinctly nonmedical grounds, which patients deserve to be treated. Medical
students are hearing what we all hear from journalists, politicians, the business community, insurers, and economists: Health care in America today

costs too much. It is not surprising that some physicians in training perceive it to be their obligation to help solve this problem.

Most citizens are not in a position to engage in any actions that would cut the cost of these expenditures. Nor would most of us willingly choose to forego a needed medical treatment on the grounds that the percentage of the GNP spent on health care hovers at around 12 percent and that is too high. If our physician informs us that we need a diagnostic or therapeutic procedure, who among us will reply: "But shouldn't I be helping to save society money?" Unlike ordinary citizens, however, physicians might be in a position to save society money, especially if they believe it to be their obligation to ration medical care at the bedside. It would be instructive to take a look at some features of a health care system that does impose an obligation on physicians to ration their patients' medical care: the British system.

The British System: A Cautionary Example

The authors of an article entitled "Why Britain Can't Afford Informed Consent"[2] point out some ethical consequences of bedside rationing. Robert Schwartz and Andrew Grubb contend that "Because the system cannot tolerate the financial effects of all reasonably possible patient decisions, it cannot tolerate the doctrine of informed consent." They quote a passage from the British Medical Association's *Handbook Of Medical Ethics:* "Within the National Health Service resources are finite, and this may restrict the freedom of the doctor to advise his patients, who will usually be unaware of this limitation. This situation infringes the ordinary relationship between patient and doctor."[3]

The feature of the British system most often noted in connection with finite resources is the absence of patient choice. What Schwartz and Grubb point out is the implications the system has for informed consent and, more generally, the doctor–patient relationship. They refer to the U.S. system of medical economics as one of "consumer sovereignty," claiming that "there can be no 'consumer sovereignty' except within a system that recognizes the doctrine of informed consent, and there is no informed consent where the patient-consumer cannot have a choice of health care options."[4] Now it is evident that anyone who is a proponent of rationing must also pay the price of rejecting the economic principle of consumer sovereignty, because the very idea of rationing is incompatible with it. Are those who urge rationing prepared to give up the ethical doctrine of informed consent? Perhaps not, but Schwartz and Grubb argue that this is a price that may have to be paid.

Their argument rests on certain features of the British system that make rationing mandated by law impossible, while matter-of-fact rationing has become the operative method within the National Health Service. The most notable features are the decentralized nature of the National Health Service "combined with the medical profession's consistent demand that it be the sole arbiter of medical questions."[5] The effect is to put the burden of allocating resources on individual physicians. This, in turn, results in physicians defining the criteria for selecting patients as narrowly medical questions, ones about which doctors alone have the requisite expertise to make decisions. Physicians are thus inevitably confronted with a built-in conflict of obligations: to their individual patients, the traditional physician's role; and to the larger society.

What happens is this: The physician treating a patient who has end-stage kidney failure is obliged to explain to the patient the risks, benefits, and alternatives to dialysis, a standard life-prolonging treatment for patients whose kidneys have ceased to function. However, the doctor also plays the role of allocator of health services. In that role, the physician is obliged to see to it that marginal patients are not provided with dialysis. The British system requires physicians to play the second role, that of allocator of scarce resources. How, then, can the physician be expected to play the role of clinician adequately, providing the patient with full information? Since the alternative to dialysis is usually death, it is hard to imagine how a physician who must play two conflicting roles simultaneously can possibly play both of them well.[6]

Physicians find it hard enough to talk about death with patients when there are no existing treatments that can prolong life. But when a life-prolonging treatment does exist, and when the physician is the person who must deny that treatment to one patient in order that the money can be better spent on another patient, few are likely to be so blunt as to tell this to the patient who is destined not to get the therapy. In such situations, the patient could hardly view the personal physician as an advocate.

Rationing by Hospital Administrators

Rationing in the clinical setting is not limited to decisions made by physicians. Rationing can also be initiated and urged by hospital administration. This can result in interference with a clinician's decisions regarding a particular patient, or it can escalate into a more general practice of urging staff to shorten hospital stays for an entire category of patients. Here are two illustrations.

In the first case, a woman about 70 years old had been brought to the emergency room with a temperature of 107°F. Seven months earlier, following her admission to the hospital, she was given a diagnosis of status epilepticus. At the time her case was brought to the ethics committee by her physician, she was in the intensive care unit, on a respirator from which she could not be weaned. Brain damage was far-reaching. It had become increasingly difficult to gain access for the insertion of intravenous lines. The patient did not respond to stimuli, had already developed large decubiti, and required regular suctioning and turning. These tasks of daily maintenance were carried out by a private duty nurse.

The patient's family—her husband and two daughters—were told of the bleak prognosis from the beginning. Yet they continued to insist that this level of care be maintained. The husband refused to consider the physician's suggestion to write a DNR for the patient. Contrary to expert medical evaluations, the family believed the patient was communicating with them. They refused to believe the prognosis and were clinging to the hope that she would recover.

There was a troubling aspect to this case, which was the reason the physician sought the help of the ethics committee. The hospital was paying for the private duty nurse and a costly bed in the intensive care unit. The patient was being given antibiotics, as well as artificial food and fluids to maintain her life. The physician came to the ethics committee in the belief that he should be an advocate for the family of his patient, and that the right course of action was to honor their wishes about the level of care being provided. Yet he was being criticized, even threatened, for refusing to make financial considerations the overriding factor. The message was clear: This patient was absorbing a disproportionate share of the hospital's resources. The doctor had been pressured by the hospital administration to discontinue this level of therapy because of financial costs.

The physician believed that the patient could not live much longer and that it would be cruel to ignore the family's request for continued treatment. It turned out that he was right about the prognosis, as the patient died soon after the case was reviewed by the ethics committee. Luckily, an ultimate confrontation between the physician and the administration had been avoided.

Note that the pressure was coming primarily from the hospital's central administration and that the underlying reason was purely financial. This must be kept clear and distinct from allocation of scarce hospital resources, such as intensive care unit beds, where the decision maker is the physician in charge of the unit, seeking to make a fair or just allocation of these resources. Deciding to remove a patient from an intensive care unit because of a genuine scarcity of resources has the effect of denying care to one patient, but with the intended and expected consequence of providing the bed to another patient who would receive greater benefit. Furthermore, when the decision maker is an individual in charge of admitting and discharging all patients in a given unit, there is at least the opportunity to employ criteria for admission and discharge in a fair and consistent manner. To be analogous, decisions by a hospital administrator to save money on a particular patient would have to result in spending the money saved on another identifiable patient who is likely to benefit more than the first patient; and to do this fairly, the administrator would have to survey the relative needs of all patients in the hospital to ensure that the criterion of shifting resources to patients likely to benefit most is being applied in a consistent and even-handed manner.

It is true that families sometimes maintain a false hope—in their grief or denial—that a miracle will occur. Physicians are normally reluctant to ignore or override the wishes of a patient's family in such cases, not only out of fear of being sued but also out of respect for the next of kin as surrogate decision makers for patients who lack decisional capacity. A sound reason for deferring to family members is that they are the ones most likely to know what the patient would have wanted regarding continuation or cessation of life support.

A less compelling reason, ethically speaking, for physicians' unwillingness to ignore or override family member's wishes is their desire not to offend family members. When the patient can no longer participate in decisions, it is the family the physician must deal with. Always lurking in the background is the fear of liability, a fear that haunts individual physicians as well as the hospital administration.

There is a more general type of bedside rationing that can be initiated by the hospital administration. This occurs when the administration regularly exerts pressure on medical and nursing staff to discharge patients. An administrator says that it's time to discharge, and the medical staff says that the patient isn't ready. In the next illustration, patients are receiving an "alternate level of care (ALOC)." They are not occupying beds in an intensive care unit or in the acute care portion of the hospital. But for patients in this category, there is lower reimbursement or, in some cases, no reimbursement at all to the hospital. The possibilities of discharge are limited to a nursing home or a skilled nursing facility; to the patient's own home; or to a shelter or the street.

Medical and nursing staff have argued that they are obligated to provide the safest form of discharge for patients. They are reluctant to force patients to leave the hospital, especially in circumstances where patients refuse at the last minute. Since beds in the ALOC unit are not usually filled, administrators cannot use the justification that they are responding to the problem of allocating scarce medical resources.

Still another form of rationing takes place at a bureaucratic level. The following example involves approvals of equipment for individual patients. The place is a medical unit in the New York City Health Department that deals with children with developmental and other disabilities. This time the pressuring agency is Medicaid, which tries to get physicians in the medical unit not to approve certain equipment for individual children because the equipment costs too much money. However, the assigned function of doctors in the medical unit is to make approvals or disapprovals on medical grounds alone. Once that review takes place, the request then goes to Medicaid to approve or disapprove on financial grounds.

Medicaid personnel have been trying to get Health Department physicians to incorporate nonmedical criteria into their medical decisions, and thus to cease approving expensive, high-technology equipment for children with disabilities. The equipment includes "power chairs," electrically powered wheelchairs; augmented communication devices; and environmental control units (ECUs) that enable severely impaired individuals to control numerous devices in their environment. If requests for equipment are disapproved based on medical criteria, those requests never reach the Medicaid office for approval or disapproval based on financial grounds. Physicians on this Health Department panel say they are being pressured by Medicaid not to issue medical approvals for these types of equipment for children who need them.

It is a common mistake to confuse expensive resources with genuinely scarce resources. The motivation for seeking a fair means of allocating scarce resources is to achieve justice in distribution. The motivation for controlling the use of expensive resources is to limit costs. If money saved by rationing costly services actually succeeded in being reallocated to other patients in greater need, that could be justified by a principle of fair allocation of resources. But the way our health care system is structured, such reallocations simply do not occur.

An explanation of why fair reallocations cannot be made was given by Arnold S. Relman, a former editor of the *New England Journal of Medicine*. Relman has argued that in this country, the systems of reimbursement for medical services, hospital financing, and private and public insurance are not set up so that saving in one place will actually result in appropriate expenditures in another.

> In a system with a fixed global health care budget established by national policy, as in the United Kingdom, physicians forced to withhold potentially useful services from their patients because of costs at least can be assured that the money saved will be appropriately spent to help other patients and that all publicly financed patients will be treated more or less alike. But this is not so in a disorganized and fragmented health care system like ours, which has multiple programs for the care of publicly subsidized patients and no fixed budget.[7]

Relman's main point is to deny the assumption that money saved by rationing in our system would succeed in being appropriately spent to help other patients.

Would Rationing Work?

A few commentators question the need for rationing medical care or the likelihood of rationing's achieving its aim: to control health care costs. Relman is one of these skeptics. He contends that "Even if a workable, medically sensible, ethically and politically acceptable rationing plan could be devised, it would not save money in the long run."[8]

In support of this claim, Relman argues that those who call for rationing in order to cut costs have misdiagnosed the causes of increased expenditures on health care. Contrary to popular belief, it is not new forms of technology and insatiable demand that are the fundamental causes of cost inflation, nor is it overuse, inefficiency, duplication, or excessive overhead expenses.[9] Instead, he argues, "It is the way we organize and fund the delivery of health care that rewards the profligate use of technology and stimulates demand for nonessential services; it is the system that allows duplication and waste of resources and produces excessive overhead costs."[10]

A revealing study confirms Relman's contention that excessive overhead costs are part of the problem of runaway health care expenditures in the United States. The authors of the study, Woolhandler and Himmelstein, remark that "A cynic viewing the uninflected curve of rising health care spending might wonder whether the cost-containment experts cost more than they contain."[11] The study found that overall expenditures for health care administration in the United States account for up to 24.1 percent of the total amount spent for health care (in contrast to 11 percent in Canada).

Even more dismaying than the amount of money devoted to administration of the system is the fact that inefficiency appears to be growing, in part as a result of the quest for efficiency itself: "Cost-containment programs predicated on stringent scrutiny of the clinical encounter have required an army of bureaucrats to eliminate modest amounts of unnecessary care. Each piece of medical terrain is meticulously inspected except that beneath the inspectors' feet."[12] One ironic result of all the attention paid in recent years to the rising costs of health care is the increase in the number of administrators and bureaucrats required to monitor the behavior of doctors, nurses, and patients—"an enlarging audience of utilization reviewers, efficiency experts, and cost managers."

Much of the data reported in this study consist of comparisons of the costs of health care administration in the United States and Canada. Perhaps the most dramatic point here is that if administrative costs in this country were reduced to Canadian levels, it would save enough money to fund coverage for all Americans who are currently uninsured and underinsured.[13]

A growing portion of these administrative costs go for the procedure known as "utilization review." This is a method of overseeing decisions made by physicians regarding medical procedures they recommend and the length of hospitalization stays for their patients. The utilization review is carried out by independent companies whose function is to keep medical costs down. The companies themselves say that they are advocates for the patient,[14] a role they presumably perform by checking physicians' treatment plans to make sure that only medically "necessary" procedures are done. The president of one utilization review company described this role by saying, "It should give the patient comfort that another group of doctors agree with what his doctor wants to do."[15] However, it could hardly give a patient comfort to learn that a group of outside doctors might choose to override the personal physician's decision to keep the patient in the hospital for a longer stay.

The chairman of a division of one company that does utilization management for many insurance companies seemed eager to give assurance that patients' interests are not being undercut by the process: "The well-defined utilization review process does not reduce care by creating barriers to care."[16] This concern for the rights and interests of patients is reflected in the company's name: the Ethix Corporation.

Many physicians object to this form of oversight. Some complain about being beleaguered by ever-increasing bureaucratic demands on their time. Others object to being second-guessed in their clinical decisions by someone sitting in a distant office who has never seen the patient. Still others are angered by having to justify their treatment recommendations after the fact or having to determine in advance just what the utilization review standards are.

The American Medical Association devised a set of policy statements that includes the following: "Any entity which imposes barriers between a physician and a patient in terms of the physician's delivery of care should be liable for harm incurred by a patient as a result of such barriers."[17] The traditional role of the American Medical Association has been to protect the autonomy and interests of the medical profession. Although that stance has often been criticized, it is also true that the interests of the medical profession need not be advanced at the expense of patients' interests. To seek to protect physicians and patients alike from outside intrusion can serve the interests of both parties, and helps to keep physicians and patients as allies.

Can physicians continue to be their patients' allies if they are called upon systematically to ration medical care? Even if that alliance were to break down, would it nevertheless be sound policy to impose on doctors an obligation to the larger society, an obligation that might be met by withholding some tests or treatments to individual patients in their care? This leads naturally to a further question: What is the physician's ethical role in rationing?

The Physician's Ethical Role in Rationing

Recent literature reveals three basic views in reply to this question.[18] Each argues for a different "proper balance" for physicians in their dual responsibility of serving as advocates for their patients and serving as agents of social responsibility. The strongest pro-patient view is called "unrestricted advocacy." According to this position, doctors have a primary duty to their individual patients, a duty that flows from their role as physicians. Only when that primary obligation is fulfilled can considerations of social justice enter into the picture for the individual clinician caring for patients.

The second view is "minimally restricted advocacy." This position allows some considerations of social justice to enter into physicians' actions because of the connection between principles of professional ethics and those governing just institutions. The third view, "maximally restricted advocacy," argues that considerations of social justice and social welfare provide a moral basis for physicians' actions other than those directed toward their individual patients. How can these different positions be justified?

The justification for the unrestricted advocacy position rests on the notion that the primary responsibility of physicians is to care for patients—providing diagnosis, treatment, and patient education. This is how most people think of their doctor's role. Although as citizens physicians may have other obligations, such as broader social and political responsibilities, those latter obligations do not justify curtailment, slackening, or termination of the activities demanded by their primary responsibility to their individual patients.

One opponent of any restrictions on physicians' advocacy, Norman Levinsky, states unequivocally that "physicians are required to do everything that they believe may benefit each patient without regard to costs or other societal considerations. In caring for an individual patient, the doctor must act solely as that patient's advocate, against the apparent interests of society as a whole, if necessary."[19] Does this mean that physicians must do everything technically possible for each patient? Does it mean that doctors are obligated to prolong life under all circumstances? Not necessarily. Levinsky argues that physicians should "consider solely the needs of their individual patients."[20] It is obvious, however, that "doing everything possible" does not always serve the needs of the individual patient. It is sometimes contrary to patients' interests to prolong their lives by all available means.

How would the principle of unrestricted advocacy apply to the case brought to the hospital ethics committee when the administration was pressuring the physician to lower the level of care while the family requested continued treatment of their brain-damaged, elderly relative? Although that patient's life was being prolonged by maintaining the more aggressive level of care, prolonging her life could not be seen as a benefit to the patient because she entirely lacked awareness of herself or her surroundings. Therefore, the principle of unrestricted advocacy would not require the physician to continue that level of care. The principle focuses on the needs and interests of patients, not on requests made by grief-stricken family members. However, if the patient had issued an advance directive expressing a prior wish to have medical treatments continued under these circumstances, then she would be wronged (even if not harmed) by a decision to forego those treatments.

The intermediate position, restricted advocacy, appeals to considerations of justice in health care. This position would make the physician's primary responsibility one of obedience to principles of just health care. According to one statement: "It is justice that should be primary here . . . and professional ethics should govern roles circumscribed by just institutions."[21] A physician who fails to serve a patient's interest fully, for example by not making certain treatments available, would not necessarily be violating professional ethics. On the contrary, professional ethical principles may require a physician to do just that.[22]

In response to this view, I contend that it is debatable whether justice requires or permits physicians to abrogate their other role responsibilities in an attempt to serve the ends of justice. It is likely to turn out that when physicians engage in rationing at the bedside, it leads to gross injustices in the services provided to patients. I am not making a categorical claim that considerations of justice are irrelevant in the clinical setting, but rather that requiring individual physicians to engage in rationing will fail to achieve the ends of justice.

In the third view, maximally restricted advocacy, the obligation of physicians is weighted even more in the direction of society at large. According to one proponent of this position, physicians have a strong duty to promote the welfare of the broader society, and this duty may override their responsibility to promote the interests of specific patients.[23] One line of reasoning in support of this view is the argument that society contributes a great deal to physicians' training and education, so physicians owe something in return. It is true, of course, that public as well as private money is used to fund professional education and the medical research on which professional practice rests. It is plausible to argue that physicians owe something to society and that reciprocity of some sort on the part of physicians is an appropriate requirement of justice.

But is restricting the ability of physicians to advocate for their individual patients the proper form that reciprocity should take? It is hard to see just how doctors' rationing at the bedside would succeed in paying society back for its financial support of the profession of medicine. Many of those very patients whose tax dollars have been used to finance medical education and research will now be old, sick, or disabled. Who is "society" if not the patients a physician is called upon to treat? A more fitting form of reciprocity would be to urge or require doctors to donate a portion of their services to the poor or uninsured, which would actually result in reducing the tax dollars used to fund specific demands for professional service.

A proponent of maximally restricted advocacy is the economist Lester Thurow, who wrote an essay entitled "Learning to Say 'No'." Using the economist's standard tool of cost-benefit analysis, Thurow urges: "Instead of stopping treatments when all benefits cease to exist, physicians must stop treatments when marginal benefits are equal to marginal costs."[24] He takes for granted that this economic decision-making tool is the most appropriate one to use in determining when to deliver medical treatments to individual patients.

What form of rationing is ethically desirable? Thurow argues against the form of rationing practiced in the British system. He rejects a central or categorical form of rationing whereby "third-party payers can write rules and regulations concerning what they will and will not pay for."[25] His objection is that such a procedure works clumsily, since rules cannot be adjusted to the nuances of individual cases. Instead, Thurow's preferred solution is for doctors to build up a "social ethic and behavioral practices that help them decide when medicine is bad medicine—not simply because it has absolutely no payoff or because it hurts the patient—but also because the costs are not justified by the marginal benefits."[26] The desired result will presumably be "a system of doctor-imposed cost controls that will be much more flexible than any system of cost controls imposed by third-party payers could be."[27] Whether the virtue of "flexibility" is likely to be accompanied by other virtues, such as justice and beneficence, Thurow leaves unexplored.

Like most people who argue that rationing is needed because too much money is being spent on health care in the United States, Thurow observes that "every dollar spent on health care is a dollar that cannot be spent on something else."[28] That proposition is, of course, trivially true. Every dollar spend on *x* is a dollar that cannot be spent on *y*. Thurow's proposed ethical remedy is to transform physicians from healers concerned to provide for their patients' medical needs into cost-benefit analysts at the bedside. It is hard to think of anything more contrary to medical ethics in clinical practice.

Ethical Arguments against Rationing at the Bedside

I conclude that even if expenditures for health care need to be cut, rationing at the bedside is an unethical way to go about that task. I cite five lines of argument in support of my contention.

1. Requiring or urging physicians to ration at the bedside undermines the advocacy role of physicians for their patients and builds a conflict of obliga-

tions directly into the physician's professional role. This is not to suggest that in the absence of turning physicians into fiscal gatekeepers, conflicts never occur. On the contrary, a number of widely recognized conflicts often occur: dilemmas of confidentiality when a patient poses a risk of harm to others; problems stemming from the fact that the twin obligations of beneficence and respect for autonomy cannot always be simultaneously fulfilled; conflicts of obligation in the clinical setting when physicians conduct placebo-controlled, randomized trials using their own patient populations; and others. Professional and personal lives are replete with all sorts of conflicts of obligation. Nevertheless, urging or requiring physicians to ration medical services at the bedside would build in a permanent structural conflict, putting doctors in a perpetual quandary over which obligation should take precedence.

2. Bedside rationing is likely to result in gross injustices in which like cases are not treated alike. If rationing is left to the physician's discretion, with doctors being exhorted to save society money because of the runaway costs of health care, an array of arbitrary and inconsistent decisions is likely to result.

Some physicians will take their obligation as society's fiscal gatekeepers very seriously and engage in rationing of the type practiced by the doctor who thought it was an abuse of Medicaid to pay for services for Lori because she had borderline intelligence and had not complied with medical orders in the past. Other physicians will adhere steadfastly to the view that their first obligation is to their patient's health and well-being. Some doctors might decide to ration based on their own subjective values regarding certain types of patients—for example, drug addicts, alcoholics, or retarded or demented individuals. The overwhelming likelihood will be a violation of justice as fairness, which requires that similar cases be treated similarly.

A more systematic form of discrimination resulting from the use of cost-benefit analysis is likely to be age discrimination. One geriatrician has argued persuasively that the use of cost-benefit analysis in geriatric care results in turning "age discrimination into health policy." This is because that economic tool depends on techniques for quantifying benefits that have a built-in bias against expenditures on health care for the elderly. Physicians would then be encouraged to withhold expensive care from patients generally on the basis of their age, even if such care is likely to benefit the individual patient greatly.[29]

Still others physicians, after taking two decades of flak from bioethicists and enduring the expansion of patients' rights at the expense of their own decision-making authority, will recognize an obligation to respect their patients' autonomy. That is a good thing for autonomy but also a danger from the standpoint of justice. Patients who exercise autonomy may have their requests for expensive therapy respected, while those who tend to relegate decision making to their physicians may be given fewer or lesser options, since these are the patients on whom society's money need not be squandered.

As for incapacitated patients, especially those with no families to advocate for them, they may well be the first candidates for rationing. If rationing is done at the physician's initiation, this category of patients poses the fewest problems, since neither patients nor their families can pose questions or raise objections. If rationing is done at the behest of the hospital administration, there need be no fear that a patient or family will sue the hospital for providing less than adequate care. There will be a natural incentive for physicians and hospitals to carry out rationing schemes with patients unable to speak for themselves and without vocal family members to serve as their advocates. Again, the result will be injustice: treating like cases differently or unlike cases similarly with regard to providing appropriate medical treatment.

3. Rationing at the bedside by physicians is not likely to achieve the desired aim: cutting costs in one place, where interventions are less beneficial from a medical standpoint, in order to shift resources to another area where patients are underserved or more needy. Consider, for example, the claim often made that if we save money on these elderly, incapacitated patients admitted to the intensive care unit, we can then provide medical services where they are more needed—say, for poor women in need of prenatal care.

Are these assumptions correct? It is surely true that more prenatal care is needed for poor women. But the way health care is financed at present in the United States, there is no systematic or efficient way of transferring funds from one sector to another—for example, from Medicare for the elderly to a neighborhood family planning clinic for teenagers. As a means of achieving a more just system for distributing health services, rationing is bound to fail. It would fail to achieve the aim of reallocation of health care funds to more appropriate areas, and it would fail to treat like cases alike.

4. Instituting rationing by personal physicians as a regular practice militates against informed consent and openness in the doctor–patient relationship. This conclusion emerges from the cautionary lesson of the British system, as described in the article by Schwartz and Grubb. If patients come to learn that their physicians perceive a duty to society that takes the form of restricting their ability to advocate for the individual patient, would it be reasonable for patients to place their full trust and confidence in their physicians? If patients believe that their doctors are doing everything reasonable and appropriate to serve their best interests, they cannot also believe that their doctors are deciding to limit costly diagnostic or therapeutic procedures that may provide some benefit. Those two beliefs could not be held simultaneously with any consistency. The British may well be happier with their doctors than Americans are with theirs. But Americans are unlikely to go

back to an earlier time in which informed consent was a rarity and paternalism by physicians the rule.

5. Ignorance on the part of physicians about a number of relevant factors
makes it almost inevitable that they will botch the job of rationing on a case-
by-case basis. This ignorance encompasses probabilities of success or failure
of treatment in individual cases and in the actual costs of various treatments.
Moreover, vagueness and uncertainty surround the concept of "low probability of success" in everyday practice: How is the practitioner to define
"low"—as a 2, 5, 10, or 20 percent likelihood of survival with a good quality
of life?[30]

A second area of ignorance is economics. Are doctors educated well
enough about microallocation of resources to perform rationing properly?
People enroll in two-year MBA programs based on the belief that such
educational programs are necessary for making sound business decisions that
rest on microeconomic precepts. It takes some expertise in economics and
accounting to carry out sophisticated cost-benefit analysis. Nothing in the
typical education of physicians equips them to perform such analysis with
precision or skill. This should not be construed as an invitation for economists to step in at the bedside to perform the cost-benefit work the physician
is unable to do.

A third area is particular facts about costs and billing procedures. Do
physicians usually know the actual cost of the diagnostic, therapeutic, or
palliative interventions they order for their patients? They may be more cost-
conscious than in the past, but even if they do know the actual financial
costs, that is only one side of the cost-benefit equation. The benefit side is
not calculated in financial terms, but rather in unquantifiable terms such as
the value to the patient of a prolonged life or marginal improvements in the
patient's medical condition or quality of life. It could not serve as a remedy
for this shortcoming to put an expert cost-benefit analyst on the health care
team. From an ethical point of view, the proper response is to reject the
appropriateness of using a cost-benefit approach in the care of individual
patients.

A different question is whether the stated costs of various treatments or
those billed to insurance companies are really the actual costs. The situation
may well be like that uncovered several years ago in the Defense Department, when it was discovered that toilet seats and coffee pots were being
purchased at a cost of several thousand dollars each. The justification offered
was that these ridiculous prices were simply a device of "accounting procedures." We may know how much hospitals charge for a one-week stay in
the neonatal unit, but do those charges reflect the actual cost of caring for a
particular premature infant? Or are they derived from a more general cost-
accounting scheme that produces large revenues for the hospital in some

areas in order to offset unavoidable losses in other areas? A physician who decides that a particular patient can probably do without a computed to-mography (CT) scan may feel virtuous by acting to save society money. But that act of rationing could result in a loss of money for the hospital or the radiology department, which is eager to maintain or increase the number of CT scans because third-party reimbursements pay for the expensive machine and thus enhance overall revenues.

Finally, rationing at the bedside requires physicians to be deceptive or at least to withhold information from their patients who are subjected to ration-ing decisions. At worst, it transforms doctors from advocates of their patients into adversaries. The bioethical principle of beneficence, obligating doctors to strive to bring about the best outcome for their patients, is in danger of being replaced by a cost-benefit mentality. And with that, medical ethics will be swallowed by economics.

Schemes to ration hospital care or medical treatment represent one way in which patients are individually or collectively removed from decision mak-ing. In the name of cost containment, different forces seek to ration medical care and limit the availability of treatments. When the items rationed are life-prolonging treatments or beds in an acute care hospital, the patient's right to decide can be equated with the right to preserve and prolong life. Today's rationing trend clashes with the ongoing efforts of people who wish to play a role in their own treatment decisions.

All this points to the erosion of patients' rights in the name of "the good of society." It is the proverbial "lifeboat ethics" situation: Some must be sacri-ficed for the good of all. But are we in the lifeboat yet? That is doubtful. The injustice that would inevitably accompany rationing schemes at the bedside should lead policymakers to seek an ethically more acceptable solution to the problem of rising costs in the health care system, a solution that would promote an alliance between physicians and patients rather than turning physicians into adversaries of their patients.

Notes

1. Ruth Macklin, *Mortal Choices: Bioethics in Today's World* (New York: Pantheon, 1987), pp. 163–64.
2. Robert Schwartz and Andrew Grubb, "Why Britain Can't Afford Informed Con-sent," *Hastings Center Report*, vol. 15 (1985), pp. 19–25.
3. Ibid., p. 23.
4. Ibid.
5. Ibid., p. 24.
6. Ibid.

7. Arnold S. Relman, "The Trouble with Rationing," *New England Journal of Medicine*, vol. 323 (1990), p. 912.

8. Ibid.

9. Ibid.

10. Ibid.

11. Steffie Woolhandler and David U. Himmelstein, "The Deteriorating Administrative Efficiency of the U.S. Health Care System," *New England Journal of Medicine*, vol. 324 (1991), p. 1253.

12. Ibid., p. 1256.

13. Ibid.

14. Leonard Sloane, "Diagnosing the Cure: Insurers Look Over Doctors' Shoulders," *New York Times* (August 31, 1991), p. 46.

15. Ibid.

16. Ibid.

17. Ibid.

18. Nancy S. Jecker, "Striking the Balance Between Public and Private Good," *Medical Ethics for the Pediatrician*, vol. 6 (1991), pp. 6, 10 11

19. Norman G. Levinsky, "The Doctor's Master," *New England Journal of Medicine*, vol. 311 (1984), p. 1573.

20. Ibid., p. 1574.

21. Jecker, "Public and Private Good," p. 6.

22. Ibid.

23. Ibid.

24. Lester Carl Thurow, "Learning to Say 'No'," *New England Journal of Medicine*, vol. 311 (1984), p. 1569.

25. Ibid.

26. Ibid.

27. Ibid.

28. Ibid.

29. Levinsky, "The Doctor's Master," p. 1574.

30. Ibid., p. 1573.

8

Medical Futility: The Limits of Patient Autonomy?

Suppose that you overheard your doctor say to another physician: "There's no point in continuing to treat this patient. Further treatment would be futile." If the doctor is talking about the patient in the next bed, you might assume that the physician is making a medical judgment. But if the doctor is referring to you, you might very well think: "It may seem futile to the doctor, but not to me!"

Despite a historical tradition of medical paternalism that still holds sway in some quarters, patient autonomy is a cornerstone of ethical decision making in today's medical practice. The notion of patient autonomy is not meant to supplant the time-honored ethical principle of beneficence, directing physicians to try to bring about the best medical outcomes. Rather, the emphasis on autonomy has sought to alter the long-established paternalism in medicine, to shift it away from physician-controlled decision making and toward increasing patient participation and self-determination in choosing a medically suitable treatment plan.

A discernible backlash of sorts can now be detected among some physicians and, perhaps more surprisingly, on the part of a growing number of ethicists. Supported by hospital administrators, they ask, with increasing frequency, whether respect for autonomy entitles patients to demand anything they wish. Concern is expressed that the concept of autonomy has been carried too far, especially in situations where patients tell their doctors to "do everything." May they (or their families) demand treatments that physicians say are medically futile? Do patients have a right to futile treatment?

An 87-year-old woman with a history of Alzheimer's disease and congestive heart failure was admitted to the hospital from a nursing home. While she was still in the emergency room, a breathing tube was inserted, and later, the tube was surgically placed. Subsequently, the patient was maintained on mechanical ventilation for irreversible respiratory failure and lapsed into a coma. The patient had no family, but a friend had visited and gave some indication that the patient did not want to be on life support. However, the friend had not been named by the patient as a surrogate, and there was no additional evidence of the patient's wishes.

The patient's doctor argued that further treatment was futile and was only prolonging her dying. He urged that the patient's current treatment protocol be abandoned in view of "the overall futility of the situation." In using the term "futility," the physician probably meant that the patient had no hope of recovery. Prolonging her life by an artificial breathing machine could not restore her mental function or even make her conscious and aware of her surroundings. Yet to maintain her on a respirator would not be futile with regard to keeping her alive. Without mechanical ventilation, she would surely die. Although this patient's quality of life had become extremely poor, does that mean that "futility" is the correct concept to apply to her situation?

Until very recently, the ethical issues in this case would have centered on the strength of the evidence concerning the patient's wishes, and also on the authority of a friend—rather than a close relative—as surrogate decision maker for the patient. From a legal and ethical standpoint, is the friend's word that the patient didn't want to be on life support sufficient? Is more solid evidence required? Is a friend who had not previously been designated as a surrogate by the patient an appropriate decision maker? These questions are still pertinent, but another, quite different concern is being voiced by doctors and hospital administrators.

The physician caring for this patient said: "We're spending a lot of resources, a lot of energy, on this patient. It is abhorrent to continue providing futile treatment to this patient. We have obligations to the larger society. This is an absolute waste of resources." Another physician present at the conference added: "At this hospital, we do futile treatment when the family requests it. But in this case, there is no family. Do we have to continue treatment?" A nurse familiar with the case said that a hospital administrator had called while the patient was still in the emergency room and instructed the physicians not to admit her to the hospital. "There will be no reimbursement, and we may not be able to get rid of her once she's here," the administrator said. The picture that emerged exemplified the hospital practice (though not official policy) to monitor patients' conditions to avoid hopeless cases. Cessation of treatment would be suggested and carried out as long as family members did not object.

The Case of Baby L

Rarely do real medical cases have only a single dimension. A case involving a severely ill infant whose parents requested that the doctors do everything to preserve her life created a furor among physicians who commented in medical journals.[1] The infant, Baby L, had been born prematurely, with poor respiratory function and a poor neurologic condition. She had numerous illnesses in the months following her birth, and at the age of 23 months was readmitted to the hospital with pneumonia and infections, requiring mechanical ventilation and cardiovascular support. The child was severely handicapped and remained at a 3-month level of development. Her mother continued to demand that everything possible be done to ensure the child's survival.

A meeting was convened to discuss the advisability of reinstituting treatment at the time of the baby's readmission to the hospital. Participants in the meeting were unanimous in agreeing that further medical intervention "was not in the best interest of the patient."[2] They contended that "further intervention would subject the child to additional pain without affecting the underlying condition or ultimate outcome."[3] The baby's mother rejected that opinion.

When physicians stated that treatment would be "futile and inhumane," they sought to refuse to continue therapy based on two separate considerations.[4] To claim that a treatment is inhumane is one thing. To contend that it is futile is another. These two reasons for refusing a family's request to continue treating their infant rest on entirely different ethical principles. Refusal on the grounds that treatment would be inhumane rests on the principle of beneficence: that physicians should strive to bring about a balance of benefits over harms.

The physician who was treating Baby L published a case report in a medical journal, co-authored by a theologian and an attorney. The report defends the refusal to treat Baby L by citing "the primary obligation of health care providers" to "Do no harm."[5] Yet it also appeals to "the physician's reasonable medical judgment"—in this case, the medical team's belief that Baby L's condition had deteriorated to the point where further treatment would be futile.[6] Although it is true that treatment would be futile in the sense that it could not reverse the child's underlying medical condition, it was surely not futile in its ability to prolong the infant's life.

A series of responses to this case report was published in the *Journal of Perinatology*. These brief commentaries are remarkable for the diversity of views held by doctors and other health professionals. They exhibit deep confusion by blurring the concept of futility, the role of economics in decisions about individual patients, and the physician's obligation to act in the patient's best interest. The responses illustrate not only that physicians do not speak with one voice, but also that conceptual confusion and factual errors pervade the current discussion of medical futility.

One commentary endorses the principle that "it is the physician, not the patient [or patient's family] who must sort out the possibilities. . . . [This] responsibility should not be shifted onto the shoulders of the patient in a misguided attempt to respect autonomy."[7] Having established the physician's decision-making authority, this commentary goes on to claim that physicians have an active obligation to ensure that the suffering resulting from treatments is outweighed by the benefit of an expected recovery from the underlying illness. That claim sounds very much like an appeal to the principle of beneficence, but a closer look reveals a crucial departure from that principle. Not all benefits to patients can be calculated in terms of recovery from an underlying illness. People with disabilities or chronic illnesses may never receive that benefit, yet they may seek treatments that prolong what for them is an acceptable quality of life. Acknowledging the difficulty of determining when life-sustaining therapy "becomes futile," the article nonetheless argues for the "medical responsibility to acknowledge futility and to recommend management programs consistent with this circumstance.[8] Despite its paternalist stance, this view remains patient centered.

In contrast, another commentary focuses on everyone except the patient. This is especially surprising given that the author is from the nursing profession, whose members frequently assert that it is they, not physicians, who are stalwart advocates for patients. The commentary questions whether Baby L's mother is aware of how keeping the infant alive is affecting her family, and "what about the care of her three other children?"[9] Does the author really believe she is more aware than the child's mother about the effects on her family? More troubling, however, is the quick shift to the resource issue: "We live with finite resources. When extensive extraordinary means are used to sustain the life of a child with extensive neurologic deficits, these resources are unavailable to others who have a reasonable chance of survival. . . . Individual health care needs must be balanced with the health care needs of the community."[10]

The position taken by this commentator rests on false suppositions at least, and at worst is symptomatic of a pernicious trend. If there had been a genuine problem of scarce resources in the intensive care unit where Baby L was treated, that circumstance would surely have been brought out as an ethically relevant factor. It is probably safe to assume that Baby L was not occupying a bed or using life-support technology that was being denied another infant with "a reasonable chance of survival."

The only other interpretation of this oft-repeated claim that resources are unavailable to others is that money saved by *not* treating Baby L would be used to save Baby M or N or O, who "have a reasonable chance of survival." But is there any evidence for this supposition? Baby L's medical expenses were covered by third-party payment.[11] Are there any grounds for believing that the insurance company *denied* reimbursement to another child (or adult) because the cost of Baby L's treatment absorbed a large share of the insurance pool? Insurance doesn't work that way. This justification for refusing Baby L's mother's request to continue treatment is both factually and morally flawed.

A directly opposite position is taken by another commentator, who asks what is the proper role played by economics in making medical decisions in individual cases. This physician strongly advocates patient-centered decision making: "Should the effect of the child's life or death on society have any effect on the decision? Should economics have any effect on the decision for a particular child? I believe the answer to the latter two questions is decidedly no."[12]

Finally, a thoughtful commentary tackles both the quality of life and futility issues. On quality of life:

> The central question is whether patients who cannot experience the richness of normal life have experiences that make continued existence from their own perspective better than no life at all. With aggressive treatment, the child [Baby L] survived and continued to require the extensive but preexisting home care for her neurodevelopmental disabilities. While a mental status of a 3-month-old is admittedly suboptimal, it is not self-evident that it implies a painful existence.[13]

And on futility:

> The concept of medical futility has less to do with the advisability of treatment than its effectiveness. If used in the strictly medical sense, meaning that the treatment would not be successful in reversing or ameliorating the life-threatening condition, the physicians were apparently wrong, since appropriate medical therapy was effective and the patient's prior existing status was restored.[14]

These commentaries on the case of Baby L represent a microcosm of a much larger discussion in the medical community and beyond. Those who engaged in the debate about Baby L asserted the following contradictory or at least conflicting views: the conviction that decision making must remain in the hands of physicians yet be governed by the best interests of the patient; an opposing view, stating that individual health care needs must be balanced with the health care needs of the community; agreement with the assessment that treatment of Baby L was medically futile; disagreement with the assessment that treatment of Baby L was futile; the judgment that continued treatment of Baby L was contrary to the child's best interests; the opposing argument that it is not self-evident that the mental status of a 3-month-old implies a painful existence; and finally, the charge that quality-of-life judgments were being masked as medical assessments of futility.

If some of these statements are true, others must surely be false. Despite the disparity of specific views enunciated in these commentaries, almost all rest their conclusions on a belief about what is in the best interest of the patient, Baby L. The physicians who argued that neurologic impairment does not imply a painful existence were suggesting that continued life could well have been in the infant's best interest. Others, who took the opposite view about continued treatment, judged that the application of advanced medical technology would prolong the infant's pain, suffering, and the act of dying. Both viewpoints derive from the same patient-centered ethical principle of beneficence; the difference lies in how they apply the principle to the case of Baby L. These commentators cannot be faulted for failure to advocate for the patient's interests, despite the fact that only one of the pairs of contradictory views about what actually constitutes this patient's best interest can be correct.

In contrast, the author who urged that "health care professionals need to help families realize that we . . . owe it to each other to wisely use scarce resources"[15] stakes out a very different role for the health professional than that of advocacy for the patient. That role is one of gatekeeper for society's interests, one whose job it is to ration care in the clinical setting. The perceived need to ration at the bedside is one of the driving forces behind the emergence of arguments to cease treatment based on medical futility.

This is shown by the remarks of another proponent of the rationing role for physicians, who seeks to have it both ways. Endorsing both the patient-centered principle of "nonmaleficence" (do no harm) and the principle of " 'distributive justice' (which compels physicians to allocate finite resources equitably)," this commentator cheered "an enthusiastic 'Hear! hear!' to the physicians who refused treatment of Baby L."[16] But he is mistaken in characterizing the principle of distributive justice as *compelling physicians* to allocate finite resources equitably. Although the principle of distributive justice requires that society's benefits and burdens be distributed to its members equitably, the principle itself is silent on the questions who is to perform the task of equitable allocations and in what settings.

The Elastic Concept of Medical Futility

It is easy to assent to the proposition that physicians are not obligated to offer or to undertake procedures that are medically futile. If an available treatment will not succeed in accomplishing its intended purpose, why even mention it? To justify this proposition, we have only to note that respect for patient autonomy entails disclosing to the patient the purpose of a recommended intervention, the procedures to be done, the risks and benefits of the treatment, and any *reasonable* alternatives. That last qualifier is important. In disclosing information to the patient, the physician is obligated to mention not *all conceivable* alternatives but only those that are deemed reasonable. But here is where the problem begins. A subjective element is introduced when a physician's own values about what it is reasonable to offer a patient determine how much information a patient is given. To complicate matters further, patients often have a different view about what is a reasonable treatment option.

After the doctor provides information, two additional steps are required in order to respect a patient's autonomy: ascertaining that the patient has understood the information and gaining the patient's permission to go ahead with the intervention. By enlisting the patient's full participation and abiding by the patient's wishes regarding treatment, the physician has fulfilled the obligation to respect the patient's autonomy. This interpretation of autonomy makes it out to be an asymmetrical concept. It permits fully informed, competent patients to *refuse* recommended treatments, but it does not give patients a co-equal right to *demand* anything they wish. If the physician deems a treatment medically futile, the ethical notion of patient autonomy would not apply to such situations.

This analysis is deceptively simple. Although it locates the proper role of autonomy in medical decision making, it fails to analyze the concept of futility. And it is precisely here that the current confusion lies. The concept of futility is vague and ambiguous, and its meaning is rarely specified precisely in circumstances in which it is invoked to justify physicians' decisions not to undertake a therapy or even to offer it to a patient. I agree that medical futility can be used as a legitimate ethical justification for rejecting a patient's request for a particular treatment. But the concept of futility I have in mind is a narrow one, much narrower than that which is often implied when this question comes up.

Although ethicists by and large agree that there is no obligation to provide futile therapies, it is also true that a variety of definitions of futility are currently being employed by ethicists as well as by physicians.[17] Therein lies the problem. Futility is an elastic concept that starts with a well-established, narrow meaning of the term. An example is the definition that appears in the New York State law on DNR orders. This definition sets up a time frame within which the treatment should hold:

" 'Medically futile' means that cardiopulmonary resuscitation will be unsuccessful in restoring cardiac and respiratory function or that the patient will experience repeated arrests in a short time period before death occurs."[18]

A slightly broader meaning, encompassing psychological benefits to the patient, is found in the Hastings Center's published "Guidelines for the Termination of Treatment." Here is the definition of "futility" stated in the "Guidelines":

In the event that the patient or surrogate requests a treatment that the responsible health care professional regards as clearly futile in achieving its physiological objective and so offering no physiological benefit to the patient, the professional has no obligation to provide it. However, the health care professional's value judgment that although a treatment will produce physiological benefit, the benefit is not sufficient to warrant the treatment, should not be used as a basis for determining a treatment to be futile. Treatment that is physiologically futile may offer psychological benefits and so may be warranted. The professional, patient, or surrogate may wish to consult with other health care professionals as well in assessing the futility of a treatment. The patient or surrogate should be at liberty to engage another health care professional. [19]

It is evident that the term "futility" is often used in a much broader sense than that stipulated in these two definitions, especially in circumstances where physicians seek to reject or override patient autonomy. At the opposite end of the spectrum from the narrow definition is an unacceptably broad sense of the term. It is not uncommon to hear futility mentioned when it means that the patient will not leave the hospital. Is that an acceptable stretching of the meaning?

I argue that it is not. One physician who asked "Must We Always Use CPR?" [20] refers to treatment as futile even if it can accomplish its physiological objective but where the patient will not survive to be discharged from the hospital. Arguing that in such circumstances it is not necessary to offer CPR, he claims that physicians need not even discuss CPR with patients if the treatment is deemed futile in his sense of the term.

Neither precision of meaning, clarity of thought, nor honesty in communication can justify this use of the term "futility." Can doctors predict with sufficient accuracy whether a patient will become well enough to leave the hospital? The proper characterization of CPR in this circumstance is not "futility" but rather "limited, short-term efficacy."

One situation in which it is common for physicians to employ a broad interpretation of futility is when patients have a fatal or terminal illness. Consider the following case report of a patient dying of AIDS. [21]

The patient, Mr. Evans, a 41-year-old male, was admitted with seizures and renal failure. He had a history of intravenous heroin use, was known to have HIV infection, and was given a clinical diagnosis of AIDS. At the time of admission, the patient was alert and oriented, and everyone agreed that he possessed full decision-making capacity. He consistently and continuously asserted that he wanted all medical care. Eventually his mental status deteriorated as a result of ongoing seizures and uremia. Dialysis was initiated, and the patient became intermittently alert and able to follow simple commands. However, respiratory deterioration soon led to intubation, and Mr. Evans became uncommunicative, responding only to deep pain. From that point on, the health care team questioned the continuation of some treatments and the initiation of others. Ear, nose, and throat physicians refused to perform a tracheostomy, noting that this procedure "would not improve the patient's quality of life." After it was suspected that the patient had acquired an infection from the dialysis catheter, physicians told the family that the treatment was of no use "in prolonging or enhancing life."

Arguments for and against overruling the patient's previously stated wishes about treatment were put forth. Among the arguments offered in favor of overruling the patient's treatment wishes was the following:

> . . . futility is a legitimate consideration in deciding whether treatment is to be withdrawn. It is not necessary to make fine distinctions about the precise probabilities of a given prognosis in order to decide that a particular condition is hopeless. It was clear to qualified professionals with long experience that Mr. Evans was not going to regain any degree of cognitive function, nor survive for an extended period. Mr. Evans's prognosis was felt to be medically certain by his health care team; many treatment decisions are taken on the basis of far weaker predictions. Hence, futility was a justifiable basis for determinations of reasonable medical care for Mr. Evans.[22]

I have been arguing that to stretch the concept of futility beyond its narrow meaning is illicit. But it is also unnecessary if the aim is to find a justification for terminating treatment. Ethically justifiable decisions to withhold or withdraw treatment can be based on grounds other than medical futility. One such basis is the use of the proportionality principle, which allows for the termination of life support if the benefits to the patient of continued life are outweighed by the burdens to the patient of the treatment or of life itself in that condition. If the benefits *to the patient* of continued treatment do not outweigh the burdens, then it is permissible to withhold or withdraw therapy. This principle is straightforward and relatively easy to apply in the case of patients who still have decisional capacity. Since the benefit of continued life to the patient is best judged by that patient, a benefit-burden analysis is an ethically appropriate and practically workable criterion. In such cases, patient autonomy remains intact.

Cases like that of Mr. Evans are much harder, however. Applying a benefit-burden analysis would appear to yield the conclusion that continued life cannot, in any reasonable sense, be a benefit to patients who are uncommunicative, unconscious, respond only to deep pain, and have no probability of regaining any cognitive function. Had Mr. Evans not made any statements about what he would want while he still retained decisional capacity, the proportionality principle could justify withdrawing or withholding treatment. But Mr. Evans did make such statements. Is it ethically permissible to ignore a patient's previously stated expressions of autonomy and switch to very different ethical grounds, the benefits-burdens criterion? Or is it ethically obligatory to honor those wishes, despite the fact that the patient cannot actually experience any benefits of continued life? Reasonable people disagree sharply on the answers to those questions. My only argument here is that it is not acceptable to fudge the solution to this ethical dilemma by broadening the meaning of "futility." That maneuver may appear to settle the ethical dilemma, but only at the price of linguistic dishonesty. The maneuver is designed to put decision-making authority in the hands of the doctors.

Broad interpretations of futility are adopted in circumstances in which physicians do not want to offer or undertake the intervention. The motive underlying the wish *not* to treat may be complex. A physician's wish not to offer a therapy may stem from the physician's subjective judgment about the patient's quality of life: "This patient's overall condition does not warrant aggressive or time-consuming or expensive medical interventions." Examples of patients about whom this judgment is made include elderly patients with irreversible dementia, patients of any age with profound neurologic damage, and patients with terminal cancer or AIDS. In other cases, a manifest judgment of a patient's social worth is involved: "This patient doesn't deserve the treatment." Although rarely stated in so explicit or bold a manner, it is the reason underlying the refusal of cardiothoracic surgeons to perform repeat valve replacements on intravenous drug users who have reinfected their new heart valve by taking drugs.

A third reason for not wanting to offer or undertake a treatment was noted earlier in discussing the case of Baby L: the perceived need to conserve expensive resources. If it is a matter of a just allocation of genuinely scarce medical resources, like beds in a critical care unit, it calls for a careful and often agonizing process of decision making. There is no need to bring in the concept of futility. It is necessary to adopt an ethically sound criterion for choosing among patients in need of the resource. However, when the resources are ones that are financially costly, like an expensive drug, or hemodialysis, or neurosurgery, then it is confused and dishonest to cloak decisions to withhold them under the mantle of futility.

The Case of Helga Wanglie

Not only has the concept of futility been broadened beyond its ordinary meaning, but physicians' insistence on stopping treatment has been expanded further to encompass "medically inappropriate" treatment. In almost every area in which ethical debate exists, worries are voiced about the slippery slope: If we take this first step, who knows where it will lead? Once a slide down a precipitous slope begins, there's no stopping the plunge. Sometimes concerns about the slippery slope are warranted; at other times, it can be shown that slides are not inevitable once a first step is taken. Regarding the use of medical futility to justify physicians' and hospitals' refusal to continue treatment, I believe the slide down the slope has already begun. A case in Minnesota that received a good deal of publicity is instructive and troubling.

Helga Wanglie was 87 years old and in a PVS. After being admitted to the hospital following a fall in which she sustained a hip fracture, her condition deteriorated to the point where she had severe and irreversible brain damage. Repeated attempts to wean Mrs. Wanglie from the respirator were unsuccessful. Her family consisted of her husband, Oliver Wanglie, also 87 years old and a retired attorney, their daughter, Ruth, age 48, who has always lived in her parents' home, and a son. Oliver and Ruth Wanglie insisted that all treatments be continued, while physicians and nurses caring for the patient at Hennepin County Medical Center in Minneapolis were trying to get them to agree to cessation of "aggressive" treatments (the hospital sought discontinuation of a respirator but not termination of antibiotics or tube feedings). After a long debate in which physicians insisted that Mrs. Wanglie would not survive an attempt to resuscitate her if she had a cardiac arrest, the family agreed to a DNR order.

The family accepted the medical diagnosis of PVS and the fact that Mrs. Wanglie was permanently dependent on a mechanical ventilator. They also accepted the physicians' judgment that she had no awareness of herself or her surroundings, including their visits to her. As is true of all troubling ethical disputes, the reasons given in support of the opposing positions are the most interesting and critical aspects of the problem.

The nurses argued that it was medically inappropriate to continue to maintain Mrs. Wanglie's life artificially. The medical director of the hospital, Dr. Michael Belzer, supported the nurses and physicians. When the Wanglie family requested that the attending physician caring for Helga be removed from the case because he kept suggesting that it was time to remove her from the ventilator, Dr. Belzer wrote in reply: "I am unaware of any physician at Hennepin County Medical Center who believes that the continued use of life-sustaining medical treatment with a ventilator is appropriate care for Ms. Wanglie at this time."[23]

The argument given by the hospital and its medical and nursing staff rests entirely on the concept of "medically appropriate care" and on the belief that only doctors and nurses (not patients or families) have the expertise to determine what is medically appropriate. Dr. Belzer made a point of noting that the hospital was not arguing that Mrs. Wanglie's respirator was needed for another patient. Moreover, the cost of her care (which by May 28, 1991, had amounted to about $800,000) was not being borne by the hospital but was being paid for by Medicare and private insurance. So, however sound the justification may be for terminating treatment for one patient because another needs those scarce resources, that justification was not used in the Wanglie case. Whether it is justifiable for a hospital to refuse to continue treatment because it costs too much is an even more dangerous slope on which to teeter.

Neither of these justifications was offered by physicians, nurses, and the Hennepin County Medical Center. Instead, Dr. Belzer wrote to the Wanglie family: "We do not believe that we are obligated to provide care that cannot medically advance the patient's personal interests."[24] Here we see the move from "medically futile" to "medically inappropriate" to "cannot medically advance the patient's personal interests." The focus has shifted from trying to determine what is "futile" to determining when a patient's "personal interests" can no longer be advanced. Still, the word "futile" was used by some physicians at the Medical Center. One physician was quoted as saying: "I know this will provide a powerful incentive for other hospitals to take a stand in cases where treatment is futile. . . . A lot of hospitals have thought of doing that . . . but it's our hospital that had the . . . moral courage and the guts to stand up and say [that] doctors should not be mandated to give clearly, overwhelmingly futile treatment."[25]

What about the Wanglie family? What arguments did they put forth to justify their request for continuing Helga's treatment? Oliver Wanglie not only claimed that he knew what his wife's wishes were, but also that they had had discussions on this subject. "We talked about these things and she stated that if anything happened to her, so that she could not take care of herself, she did not want anybody to do anything that would shorten her life."[26] Helga's daughter, Ruth Wanglie, also reported having had conversations with her mother, and recalled her mother as saying, "she wouldn't want her life prematurely shortened if she wasn't able to take care of herself."[27] Beyond these conversations reported by family members, Mr. Wanglie contended that being married to Mrs. Wanglie for 53 years enabled him to know her values. The Wanglies were a religious family, members of the Lutheran church. Mr. Wanglie said: "I'm a pro-lifer. I take the position that human life is sacred." He said his wife, "the preacher's daughter," felt the same way.[28]

Helga Wanglie never put anything in writing. The statements by her family about her values and previously expressed wishes probably do not constitute "clear and convincing evidence," in a strictly legal sense, of what Helga herself would want. But unless the family's truthfulness is questioned, these statements surely provide some evidence of the patient's values and previously expressed wishes. If evidence of that sort should be countenanced and a patient's wishes respected when a family requests cessation of treatment, shouldn't the same principle apply when families request continuation of treatment? The question is whether the principle—respect for autonomy—should be extended to cover situations in which patients have lost the capacity to speak for themselves and there is reason to believe that they would want treatment continued.

Two physicians from Hennepin County Medical Center made a number of observations on this point. Dr. Ronald E. Cranford, a neurologist and ethicist who is well known for his support of the patient's right to refuse treatment and his active involvement in cases in which families seek to have life support removed from comatose patients, questioned the sincerity or the accuracy of the Wanglie family's recollections. Cranford notes that before December 1990, Mr. Wanglie "told hospital staff that his wife had not discussed these issues, and that her views were a 'black box.' "[29] Cranford agreed that the hospital should fight to have the respirator removed against the wishes of the patient's family.[30]

Cranford's observation questioning the family's word was seconded by Dr. Steven H. Miles, also an active contributor to the field of bioethics, who served as the representative of Hennepin County Hospital's ethics committee. Miles wrote that "despite many opportunities, the family did not mention any specific views of the patient about the future use of life-sustaining treatments."[31] However, after a conference with hospital staff on December 3, 1990, Mr. Wanglie wrote a letter in which he reported his wife's alleged statements about not wanting anything done to prematurely shorten her life.[32] The implication here is that the December 3 conference led Mr. Wanglie to recognize the potential importance of his wife's own views about treatment, and thus to introduce these "recollections" at this time.

Miles reported that after the hospital board granted permission to the medical director to go to court, Mr. Wanglie cited "extensive conversations in which the patient allegedly expressed her preference for life-sustaining treatment," but added that there was no documented evidence "that this women's views are different from [those of] the vast majority of persons who would not want to be sustained if permanently unconscious."[33] In that remark, Miles adopts the "reasonable person" standard for determining what a particular patient would have wanted, moving away from a *subjective* standard, which relies on evidence gleaned from that patient's written or oral statements, her family's testimony, or, more generally, from the patient's past behavior.

Miles rejected the notion that the family should always speak for the patient, viewing that position as overly simplistic. He said that if the treatment were appropriate, the hospital would provide it. "We're trying to provide good medical care for our patients. You can't reduce medical decision-making to some incantation such as 'Mom would want.' "[34] This last statement makes it clear that in Miles's view, families should not be viewed as surrogates for patients when their requests differ from what physicians judge to be good medical care. Miles articulated that view more fully when he wrote: "The proposal that the intellectual construct of 'substituted judgment' justified by 'respect for autonomy' infinitely empowers a family over a reasoned medical conclusion that a treatment cannot serve the patient's interests defies experience and common sense."[35]

Dr. Belzer, the hospital's medical director, asked the Wanglie family to petition a court to mandate continuation of therapy. When the family declined to file a petition, the hospital filed papers with the Fourth Judicial District Court in Hennepin County. The hospital asked the court to appoint an independent conservator to make medical decisions for Helga Wanglie in the hope that the conservator would agree with the physician's desire to remove the respirator.

There is a supreme irony in this situation. Less than two decades ago, physicians and hospitals were routinely refusing to comply with the requests of families to remove respirators and other life supports from comatose patients. Those who filed petitions sought to have life supports removed from relatives or friends, while hospitals and physicians maintained that their obligation was to preserve and prolong human life, not to terminate it. Now physicians and hospitals are seeking to terminate life, using as justification their obligation to provide medically appropriate treatment.

Steven Miles sees the features of the Wanglie case as "compatible with the premises of existing 'right to die' policy and ethics."[36] The features Miles cites are the permissibility of withdrawing life-sustaining treatment if "its burdens are disproportionate to the benefits it can achieve."[37] Miles focuses on physicians' judgments concerning what is medically indicated, the criteria for personal medical interests, and what is to count as medically appropriate treatment.

In contrast, Paul Armstrong, who had been the attorney for the family of Karen Ann Quinlan in the mid-1970s, emphasizes the role of patients and their families in making these decisions. The Quinlan case, decided by the New Jersey Supreme Court in March 1976, established a precedent for a long line of cases granting patients and their families, rather than doctors and hospitals, the right to make decisions regarding termination of treatment.

Can it be that what counts as medically appropriate treatment has changed drastically in the last 15 years? Possibly, but it is more likely that doctors are still striving to maintain control over medical decision making, even if it means opposing the wishes of family members or challenging the relevance or appropriateness of directives issued by patients themselves.

The dispute over Helga Wanglie's respirator is not a technical medical dispute, but rather a dispute over what kind of life is worth prolonging. I agree with the observation that it is "presumptuous and *ethically* inappropriate for doctors to suppose that their professional expertise qualifies them to know what kind of life is worth prolonging."[38] A judgment that the lives of patients in a PVS are not worth prolonging is not derived from medical knowledge, but from values concerning what kinds of life are worth preserving. Miles is probably correct in surmising that the vast majority of persons would not want their lives to be sustained if they were permanently unconscious. Still, a small minority would want to be sustained that way. Neither of these opposing views rests on or requires medical expertise.

The basis cited by Hennepin County Medical Center and its physicians for determining what treatment is appropriate is that which "advances the patient's personal medical interests." This criterion is incoherent in this context. Helga Wanglie certainly had personal interests before she lapsed into a PVS. One of those interests, at least according to her husband's view, was in not having her life shortened prematurely. People have an interest in having their previously stated wishes honored. On the other hand and in a different sense, once Helga Wanglie fell into a PVS, she no longer had any interests— certainly none that could be fulfilled by doctors and hospitals. Patients who are permanently unconscious, who cannot experience pleasure or even feel pain, can be said not to have any interests at all.

In using a proportionality criterion, which rests on a benefit-burden analysis conducted from the patient's point of view, Miles appeals to a leading principle of bioethics. The principle itself is unassailable, but like any ethical principle, its application is what counts most in an analysis of a particular case. Miles states that "persons should not be treated for the benefit of others,"[39] thus assuming that the benefits of keeping Helga Wanglie alive accrue only to her family and not to her. That is surely true if we accept the notion that in a PVS, a person can no longer be said to have any interests and so cannot benefit from continued treatment.

But then, neither could Helga Wanglie be said to suffer the burdens of continued treatment. When benefits are weighed against burdens in appealing to the proportionality principle, both the benefits *and* the burdens must be patient centered. Miles insists that the benefits must accrue to the patient. Who, then, should be seen as shouldering the burdens of Helga Wanglie's continued treatment? Miles denies that it is the hospital or the board of directors ("the cost of this woman's care is borne by private insurance and is not an issue for the clinicians or the county board"[40]).

It appears, then, that those who suffered the "burdens" of continuing to sustain Helga Wanglie on a respirator were the physicians and other caregivers, and also, for some unexplained reason, the hospital administration. Miles points out that they were "unaware of any HCMC [Hennepin County Medical Center] physician who believes that the use of the respirator is medically appropriate" in this case.[41] He gives no other clue about what he believes to be the burdens of continuing to treat Helga Wanglie. If it is not the physicians' feelings about loss of power or authority, the only remaining candidate is the costs to society. In that case, we are back to the debate over the ethics of rationing care at the bedside in order to save society money.

There is little hard evidence for the view that the Wanglie case was triggered by the desire of physicians to regain decision-making authority that has been eroded by respect for patient autonomy and deference to the wishes of patients' families over the past 15 years or so. Yet the comments cited above made by Hennepin County Hospital's medical director, Dr. Michael Belzer, and by other doctors at the hospital are telling. Their position is strongly endorsed by another physician, a director of surgical intensive care at a hospital in Kentucky, who commented on the Wanglie case:

> The petition for judicial relief of a professionally morally oppressive request by Mrs. Wanglie's family properly belongs in a court of law. It is time to establish the limits of patients' autonomy to demand health care that the medical profession believes serves no benefit. . . . These issues are so fundamental that the ruling in Wanglie will usher in a new chapter in our efforts to respect the individual freedom of both patients and their health care providers.[42]

I believe there should be a presumption in favor of life when a patient's views are unclear or unknown.[43] That presumption requires evidence of the specific views of that particular patient. A different principle, the reasonable person standard, would allow life support to be withdrawn based on an assessment of what most people would want in such a situation. No one in the Wanglie case claimed that the patient ever said she would prefer *not* to be kept alive, despite the fact that she had remained on the respirator for several months while still conscious.[44] It was the doctors and the hospital who decided that her life was not worth prolonging. The hospital went to court to seek an independent conservator in the hope of overriding the wishes of the family.

On July 1, 1991, Judge Patricia Belois of Hennepin County Court in Minneapolis rejected the request by physicians to appoint a guardian other than Oliver Wanglie to make medical decisions for his wife. The judge's opinion began by noting that "The Court is asked whether it is in the best interest of an elderly woman who is comatose, gravely ill, and ventilator-dependent to have decisions about her medical care made by her husband of 53 years or by a stranger." The opinion went on to note that "except for unconvincing testimony from some physicians and health care providers" at the hospital, there was no evidence that Mr. Wanglie was unable to perform the duties of a guardian. The judge found Oliver Wanglie to be "in the best position to investigate and act on Helga Wanglie's conscientious, religious and moral beliefs." The judge concluded that it was in her "best interest" for her husband to be appointed her guardian, and further, that Oliver Wanglie was "the most suitable and best qualified person" to carry out those responsibilities.[45] Helga Wanglie died shortly thereafter.

Proper Application of the Concept of Futility

Physicians should not automatically be faulted for using the term "futility" in an elastic or inconsistent manner, given the confusion and lack of precision in the medical literature surrounding this term. Since "futility" is a non-quantitative expression of probability, it means different things to different doctors. "Some physicians may only invoke futility if the success rate is 0%, whereas others invoke futility for treatments with success rates as high as 13%."[46] Although some may disagree, I think it is important to distinguish between "no-yield medicine" and "low-yield medicine."

Despite the confused and conflicting usage of the concept of futility, there are now some encouraging signs that thoughtful physicians are struggling to develop a correct definition and proper application of the term. One example is a letter to the editor published in the *Journal of the American Medical Association,* in which a physician states that he recently recommended to the state of Colorado's peer review organization "that hospital DNR policies be changed to allow the physician to write the DNR order without the patient's permission when CPR is believed to be futile."[47] The letter specifies a precise meaning of "futility," which comports with the narrow definition I have been urging: "Cardiopulmonary resuscitation is futile when one or many diagnoses cause a deterioration of physiological functioning such that death is both imminent and unavoidable; that is, the cardiopulmonary arrest that will inevitably occur cannot be reversed."[48] Furthermore, from a medical standpoint, the patient's diagnosis alone should not be used as a prognostic indicator of futility, but measures of physiological deterioration should be employed since they are better predictors.

Another physician adds an important dimension, arguing that "once physicians have determined that CPR is futile and will not be performed, they should so inform patients and families."[49] Noting that doctors often "cave in to unreasonable demands for futile interventions,"[50] he makes a plea for honesty and disclosure, claiming that communication and feedback should be encouraged at a time when they are essential. This approach does not tackle the problem of medical futility by asserting that doctors must wrest control back from patients. Rather, it takes the more thoughtful tack that honest communication with patients and families can simultaneously serve the patient's best interest and embody respect for autonomy.

For reasons of conceptual clarity and linguistic honesty, among others, I have been urging that the concept of futility be understood in a narrow sense.

Moreover, when clearly spelled out, this narrow definition is easier to apply consistently in clinical practice than any of the broader, vague senses discussed earlier. The claim that judgments of futility require *medical* expertise is true only when the narrow meaning is employed. I maintain that it is ethically wrong to broaden its meaning for the purpose of justifying medical decisions that serve to override patient autonomy. Futility should be a patient-centered concept. Physicians' judgments should be limited to the following two necessary conditions for applying the concept, conditions that do require medical expertise:

1. Medical indications: an objective element, based on the patient's diagnostic category, what the medical literature says about the condition, and results of the patient's physical findings, including laboratory results and clinical examination.
2. Clinical evaluation: carried out by a competent physician and based on the medical indications, as just defined, including the patient's medical history and the clinician's assessment of physiological deterioration.

As for the decision on whether or not to initiate or continue treatment, that decision must include a third element: the competent patient's wishes about treatment. The patient need not be a health professional to express these wishes, since they incorporate the ordinary person's perception of psychological benefit. To the alert and oriented patient, a few weeks or even days of continued life may constitute a benefit. The importance of the patient's beliefs about futility are illustrated by the meaningfulness of a patient's reply to a doctor's statement that a particular treatment would be futile: "It may be futile to you, doc, but not to me."[51] The inclusion of psychological benefit to the patient in the definition of futility in the Hastings Center's "Guidelines on Termination of Treatment" would apply to this situation.

In the case of patients who still have decisional capacity, it is possible to arrive at the ethically acceptable use of the concept of futility consistent with patient autonomy. If continued or contemplated treatment would indeed be futile in the above-defined sense, physicians may ethically refuse to comply with a patient's request, but they are still obligated to communicate with the patient about their evaluation and their decision not to treat.

Matters become murkier once a patient has lost mental capacity. May physicians then use the concept of futility to justify cessation of treatment when a patient's family requests that treatment be continued? The same considerations used to analyze the situation of patients with decisional capacity apply to cases in which patients no longer have that capacity. The issue of overriding the patient's autonomy will arise only for incapacitated patients who had made their wishes known before losing capacity. Physicians might try to stretch the concept of medical futility, using one of the broader meanings to justify their refusal to treat. But if the patient has expressed a prior wish, as Helga Wanglie's family says she did, treatment may not be withheld.

In the case of incapacitated patients who never expressed any wishes about treatment and whose families request that treatment be continued, consistency calls for employing the same narrow definition of futility. But if physicians or hospital administrators seek to override the family's request for treatment, refusal to comply with the family's demand would *not* violate the principle of respect for autonomy. Properly understood, that principle refers to the autonomy of *patients*, including an extension of their autonomy through the mechanism of an advanced directive (a living will or health care proxy). To override the wishes of the family of an incapacitated patient who had made no advance directives would go against a long tradition of honoring the wishes of a patient's family but would not be a violation of any well-articulated ethical principle.

In order to be just, such decisions would have to flow from a consistent policy. Otherwise, the result would be unjust because, in some cases, a family's request to continue treatment would be honored and in other cases overridden. Although there is no clearly articulated ethical principle that compels adherence to a family's wishes (in the absence of any information about what the patient would have wanted), justice requires that like cases be treated alike.

In the words of one physician-ethicist: "We must be clear by what we mean by futile. A treatment is futile if it will not succeed in reversing the specific physiologic problem for which it is initiated. It is important not to misuse the language of futility to mask quality-of-life judgments."[52] Honesty demands that the issue of quality of life be confronted squarely. Although it is not ethically controversial to maintain that genuinely futile therapy may be discontinued, ethical uncertainty pervades the question of whether treatment may be stopped based on a physician's judgment of the patient's quality of life.

Notes

1. John J. Paris, Robert K. Crone, and Frank Reardon, "Occasional Notes: Physicians' Refusal of Requested Treatment," *New England Journal of Medicine*, vol. 322 (1990), pp. 1012–15.

2. Ibid., p. 1013.

3. Ibid.

4. Ibid., p. 1014.

5. Ibid.

6. Ibid.

7. David K. Stevenson, William E. Benitz, William D. Rhine, and Ronald L. Ariag-

no, "Physicians' Refusal of Requested Treatment," *Journal of Perinatology*, vol. X (1990), p. 408.

8. Ibid., pp. 408–9.

9. Barbara Derwinski-Robinson, ibid., p. 409.

10. Ibid.

11. Paris et al., "Occasional Notes," p. 1014.

12. Richard L. Schreiner, "Physicians' Refusal of Requested Treatment," p. 412.

13. Robert H. Perelman and Norman C. Fost, ibid., p. 414.

14. Ibid.

15. Derwinski-Robinson, p. 409.

16. Jay P. Goldsmith, ibid., p. 415.

17. Steven H. Miles, Letter to the Editor, *Law, Medicine & Health Care*, vol. 18 (Winter 1990), p. 425.

18. Public Health Law 2961 (April 1, 1922).

19. The Hastings Center, "Guidelines on the Termination of Life-Sustaining Treatment and the Care of the Dying" (Briarcliff Manor, NY: The Hastings Center, 1987), p. 32.

20. Leslie J. Blackhall, "Must We Always Use CPR?" *New England Journal of Medicine*, vol. 317 (1987), pp. 1281–85.

21. Jay Alexander Gold, Daniel F. Jablonski, Paul J. Christensen, Robyn S. Shapiro, and David L. Schiedermayer, "Is There a Right to Futile Treatment? The Case of a Dying Patient with AIDS," *The Journal of Clinical Ethics*, vol. 1 (1990), pp. 19–23.

22. Ibid., p 21.

23. B.D. Colen, "Fight Over Life," Discovery, *Newsday* (January 29, 1991), p. 65.

24. Ibid.

25. Quoted in Colen, "Fight Over Life," p. 64.

26. Ibid., p. 64.

27. Ibid.

28. Ibid.

29. Ronald E. Cranford, "Helga Wanglie's Ventilator," *Hastings Center Report*, vol. 21 (1991), p. 23.

30. Colen, "Fight Over Life," p. 64.

31. Miles, Letter to the Editor, p. 424.

32. Cranford, "Helga Wanglie's Ventilator," p. 23.

33. Miles, Letter to the Editor, p. 424.

34. Colen, "Fight Over Life," p. 65.

35. Miles, Letter to the Editor, p. 425.

36. Ibid., p. 424.

37. Ibid.

38. Felicia Ackerman, "The Significance of a Wish," *Hastings Center Report*, vol. 21 (1991), p. 28.

39. Miles, Letter to the Editor, p. 424.

40. Ibid.

41. Ibid.

42. Michael A. Rie, "The Limits of a Wish," *Hastings Center Report*, vol. 21 (1991), p. 27.

43. Ackerman, "Significance," p. 29.

44. Ibid.

45. *In re: The Conservatorship of Helga M. Wanglie*, State of Minnesota District Court, Probate Court Division, County of Hennepin, Findings of Fact, Conclusions of Law and Order.

46. John D. Lantos, Peter A. Singer, Robert M. Walker, Gregory P. Gramelspacher, Gary R. Shapiro, Miguel A. Sanchez-Gonzalez, Carol B. Stocking, Steven H. Miles, and Mark Siegler, "The Illusion of Futility in Clinical Practice," *The American Journal of Medicine*, vol. 81 (1989), p. 82.

47. Grant E. Steffen, Letter to the Editor, *Journal of the American Medical Association*, vol. 265 (1991), p. 354.

48. Ibid.

49. Stuart J. Youngner, "Futility in Context," *Journal of the American Medical Association*, vol. 264 (1990), p. 1296.

50. Ibid., p. 1295.

51. In formulating this analysis, I have benefited from discussions with Jonathan D. Moreno.

52. Alan R. Fleischman, "Physicians' Refusal of Requested Treatment," p. 408.

9

Patient Autonomy, Physician-Assisted Suicide, and Euthanasia

What may patients legitimately demand of their doctors? At one extreme are patients who demand that their physicians do everything, including providing futile treatments. At the other extreme are patients who not only insist that treatment be withheld or withdrawn, but request that their doctors take the next step and provide them with aid in dying. Does respect for a patient's autonomy mean that the physician is obligated to assist the patient in committing suicide? A supreme test of the limits of autonomy arises when patients request their physicians to assist them in ending their lives.

Most physicians recoil when they hear the word "euthanasia" mentioned. Their negative reaction and refusal to participate in actively terminating patients' lives stems in part from their medical training, in part from a gut-level response, and in part from fear of legal repercussions. As an example of the official position taken by the medical profession, the State Medical Society of New York has a policy stating that "the use of euthanasia is not in the province of the physician." The current opinions of the Council on Ethical and Judicial Affairs of the American Medical Association are unequivocal in stating that "a physician may alleviate pain and cease or omit treatment but *should not intentionally cause* death."[1]

Some doctors who are adamantly opposed to euthanasia cite the ancient Hippocratic tradition. This is the reason given by Leon R. Kass, a physician by training who is Professor of the Liberal Arts of Human Biology at the University of Chicago. Kass identifies fixed, firm, and nonnegotiable "outer limits" for medicine set forth in the Hippocratic Oath, one of which is "no dispensing of deadly drugs."[2] He cites the promise of self-restraint in the Hippocratic Oath, medicine's "primary taboo": "I will neither give a deadly drug to anybody if asked for it, nor will I make a suggestion to this effect."[3] Kass emphasizes that he is not arguing against the cessation of medical treatment when such treatment merely prolongs painful dying, nor against the use of means of relieving suffering that may increase the risk of dying. His arguments are directed against the intentional killing of patients by physicians.

Like many other doctors, Kass draws a sharp line between withholding or withdrawing medical treatments and intentionally causing the death of a patient by active means.

We are led to ponder how the action of administering a lethal injection or providing a dose of barbiturates to a patient differs ethically from the action of removing a respirator or a feeding tube. Many physicians and nurses accept and participate in removing life support from patients. Regardless of whether unplugging a respirator, turning off a dopamine drip, or pulling out a feeding tube is judged to be active or passive means, these health care professionals recognize that they are hastening death by performing these actions. Why, then, do they resist or even deplore the action of supplying deadly pills or administering a lethal injection?

The answer lies in physicians' emotional and psychological reactions. There is a deep-seated feeling that if they provide the means that causes the patient's death, they are killing the patient. But if they remove a life support, it is the disease that kills the patient. Interestingly, this is exactly the same contention that is made by health professionals who see a moral distinction between withholding life support and withdrawing it. Their argument is that if they fail to institute life-prolonging treatment, it is the disease that kills the patient; but if life supports already in place are withdrawn, it is the doctor who kills the patient. Yet the distinction seems more plausible when applied to withdrawing-killing than it does for withholding-withdrawing. The cause of death is the same—lack of sufficient oxygen, kidney failure, dehydration, or overwhelming infection—whether physicians fail to initiate those treatments or, instead, withdraw them once begun. The case is different, however, when the physician supplies or administers a substance that otherwise would not be the cause of death. From an ethical point of view, does that matter?

Each year when I teach a bioethics seminar for medical students, I present two similar scenarios that differ only in one feature. In both cases, a competent, nondepressed patient is suffering from an incurable disease. In both cases, the patient requests the physician to assist in hastening death. In neither case is the patient's life being sustained by life support measures. We assume an identical diagnosis, prognosis, and degree of suffering in both patients. However, in the first case, the patient is able to take the lethal substance orally and the physician assists in the suicide by supplying the substance; in the second case, the patient's arms are paralyzed, requiring the physician to perform active euthanasia by injecting the substance into the patient.

Some students believe that both actions are morally wrong. Others think that both are morally acceptable. But a significant number say that the physician's action is ethically permissible in the first case but ethically prohibited in the second. What's the difference? The students argue that in the first case, the patient could, at the last minute, change her mind and not swallow the pills. I then remind them that in the second scenario as well as the first, the patient is fully lucid and mentally competent. As the doctor holds the syringe poised at the patient's arm, the physician asks: "Are you certain you wish to go through with this?" As in the first case, the patient can change his mind at the last minute. Both patients are able to stop the process at any time. What's the difference?

I believe there is no difference from an ethical point of view. In both cases, the physician's intention is to relieve the patient's suffering, and the physician's action is performed in response to the patient's voluntary request. Whether the physician is causing the patient's death directly or causing it indirectly by enabling the patient to perform the final act matters little. In neither case would the patient be able to act without the physician's assistance.

An additional point could be made: that a physician who agrees to supply pills to a patient who is capable of swallowing them, but refuses to administer an injection to a paraplegic patient, is engaging in unjust discrimination. As long as patients retain the ability to use their hands and arms and to swallow, they are eligible for physicians' assistance in committing suicide. But once they lose the ability to perform the final act with their own hands, they become ineligible. The injustice lies in allowing for physicians' assistance in helping one group of suffering patients to die but not permitting such assistance for the second group.

Proponents of active euthanasia suggest that active killing is sometimes more humane, and therefore more ethically acceptable, than allowing patients to die. When given under limited, carefully specified circumstances and accompanied by adequate safeguards, a painless lethal injection is more humane, they argue, than letting the cancer patient die a slow, agonizing death.

Two Recent Cases

Two widely publicized events in which physicians assisted the suicides of patients who requested aid in dying come under the heading of physician-assisted suicide. The two cases provide a nice contrast, demonstrating the fuzzy yet recognizable line between ethically acceptable and ethically impermissible behavior by physicians.

In the first case, a 62-year-old retired pathologist in Michigan, Dr. Jack Kevorkian, invented a bizarre contraption with which a patient hooked up to the machine could push a button activating an injection of lethal drugs. In June 1990, the physician used the invention for the first time. The patient,

Janet Adkins, was a 54-year-old woman with Alzheimer's disease who had traveled from Portland, Oregon, for the purpose of using Dr. Kevorkian's suicide machine. Mrs. Adkins was a member of the Hemlock Society, had gotten in touch with Dr. Kevorkian before traveling to Michigan, and met him for the first time in a restaurant near his home the weekend before she used his suicide machine. The actual episode took place in the back of a 1968 van in Groveland Oak County Park. Dr. Kevorkian said that what he did was not murder, because it was not he but the patient who administered the fatal drugs. After lengthy investigations and a criminal indictment of Dr. Kevorkian, a Michigan judge dismissed the homicide charges against the physician.

Although Janet Adkins had been diagnosed with Alzheimer's disease and had begun to suffer memory loss that made her unable to play the piano any longer, she remained physically vigorous until the end of her life. One of her sons reported that she had beaten him in a tennis match the week before her suicide. Mrs. Adkins made a videotape a few days before she used Dr. Kevorkian's device, explaining her decision. She also left a suicide note, which said that she did not "want to let [her Alzheimer's disease] progress any further" or to put her family or herself "through the agony of this terrible disease."[4]

Dr. Kevorkian defended his action with Janet Adkins. In direct opposition to the assertions made by Dr. Kass, Dr. Kevorkian claimed that his support for euthanasia was within the "true Hippocratic tradition":

> What I'm doing was absolutely ethical in Hippocrates' day and why he practiced assisting suicides. Doctors in those days would listen to a patient's complaint. If in the doctor's opinion the patient needed the service, he would go ahead and give the patient poison, and this was entirely ethical.[5]

Janet Adkins's right to commit suicide is ethically defensible. Dr. Kevorkian's assisting her was not. Several features of the case made his actions reprehensible. First, he barely knew the woman, having met her for the first time the previous weekend. An ethical requirement for any physician-assisted suicide is an intimate knowledge by the doctor of the patient's medical condition, his or her medical history and prognosis, and his or her mental state at the time of the patient's request and final action. Second, Dr. Kevorkian acted alone, without independent examination by at least one and preferably two other physicians who could attest to the patient's physical and mental condition. Third, he agreed to the act in a rather precipitous manner. It was not preceded by repeated assessments of the patient's persistent wish to die and observations over a period of time. These and perhaps additional safeguards are required before a physician's action of assisting a patient in a suicide can be ethically acceptable.

Furthermore, Janet Adkins's physical and mental condition at the time of her suicide made it at least ethically problematic and, in my view, ethically wrong for a physician to collaborate in her death. Not only was Mrs. Adkins not in the process of dying or anywhere near being terminally ill; beating her son at tennis the previous week and traveling alone from Oregon to Michigan demonstrated her vigor and lack of mental deterioration at that point. Physicians should be able to say truthfully that a patient is genuinely suffering (rather than merely anticipating future suffering) before they assist in bringing about a patient's death.

Finally, from an aesthetic standpoint, the location and circumstances surrounding Dr. Kevorkian's use of the suicide device gave a bizarre cast to the entire episode. These are not ethical objections, since no substantive or procedural principles are violated by carrying out a suicide in an old van in a parking lot. But the surroundings robbed the action of dignity, a value often cherished by people who choose to die sooner rather than experience a process of deterioration and loss of self. In that regard, an observation made by a physician-ethicist in Michigan is telling:

> We thought patients would be horrified by the rusty van in the parking lot, the unsterile solutions—it sounded so sleazy. Instead, a significant number want to erect a statue to the man. It became clear that many people see doctors as the *enemy* when it comes to death and dying and you have to see that as a terrible failing.[6]

The report of a second highly publicized case of physician-assisted suicide was published by a physician in the *New England Journal of Medicine* in March 1991.[7] The author, Dr. Timothy Quill, described his feelings and actions over a period of months in treating a 45-year-old patient who was dying of leukemia. Dr. Quill had known the patient, a woman he referred to as "Diane" in the article, for eight years. Albeit reluctantly, Dr. Quill supported his patient's refusal of aggressive therapy for her disease. He wrote: "I gradually understood the decision from her perspective and became convinced that it was the right decision for her."[8]

Dr. Quill described Diane as a woman for whom it was "extraordinarily important to maintain control of herself and her own dignity during the time remaining to her. . . . When the time came, she wanted to take her life in the least painful way possible."[9] The physician referred his patient to the Hemlock Society, and she subsequently telephoned him requesting barbiturates for sleep, the chief method endorsed by the Hemlock Society and recommended by its director, Derek Humphry. Dr. Quill held long conversations with Diane, ascertained that she was not despondent and that her family was aware of what she was doing, and made certain that she knew how to use the barbiturates for sleep and also the amount needed to commit suicide. He describes writing the prescription "with an uneasy feeling about the boundaries [he] was exploring—spiritual, legal, professional, and personal."[10] When the time eventually arrived, doctor and patient met again. It was clear that she knew what she was doing. Two days later, her husband called the physician to report that after saying her final goodbyes to the family, Diane had died.

Dr. Quill reported the cause of death to the medical examiner as "acute leukemia." He noted in the article that although that was the truth, "it was not the whole story."[11] He wanted to protect himself, Diane's family, and Diane herself from an invasion into her past and her body. But in signing his name to the article in one of the most prominent and widely read medical journals, the physician opened the door to an investigation that would most likely follow. In July 1991 a grand jury in Rochester, New York, declined to indict Dr. Quill for murder. Subsequently, the New York State Health Department said it would not bring charges of professional misconduct against Dr. Quill, despite the policy of the State Medical Society that the use of euthanasia is not in the province of the physician.

Many physicians who wrote articles or were quoted following the Kevorkian and Quill episodes came out strongly against active euthanasia. Yet others have said that Dr. Quill acted courageously and acknowledged that physician-assisted suicide probably occurs more often than is recognized or admitted.

Among those who believe that active killing can be ethically permissible are physicians who practice active euthanasia in the Netherlands, along with their supporters, and members of the Hemlock Society in the United States. The director of the Hemlock Society, Derek Humphry, wrote a book entitled *Final Exit*, which rose to the top of best-seller lists in 1991. The book endorses active euthanasia by physicians, suicide assisted by physicians or by relatives and friends of a suffering patient, and legislation that would render physicians who helped patients to die immune from criminal prosecution. Mainly, however, the book provides detailed information regarding effective ways to commit suicide, and suggests that people who want to prepare themselves for future contingencies begin to plan ahead and approach their physician for prescription drugs well in advance.

Humphry endorses only *voluntary* forms of euthanasia, requested by the patient, and proposes numerous safeguards to guard against abuses. The book is not entirely clear about whether the patient must be terminally ill or in the process of dying, or whether assisted suicide is permissible in cases of patients with a horrible illness or total disability who are not likely to die soon from that condition.

Supporters of legalizing euthanasia frequently point to the Netherlands, where professional and judicial guidelines permit euthanasia but limit the practice to people who explicitly and repeatedly ask for it. Opponents of legalization cite studies providing evidence that departures from these guidelines do exist, specifically that some Dutch patients are being euthanized without their consent. In this controversy, each side disputes various claims and evidence presented by the other. At least part of the disagreement stems from uncertainty and confusion over what properly counts as euthanasia.

The Ambiguous Concept of Euthanasia[12]

There is no single meaning of "euthanasia," but there is a core concept common to a number of definitions. The following are only a few of the definitions that have been offered:

1. "the painless inducement of a quick death,"
2. "an act or practice of painlessly putting to death persons suffering from incurable conditions or diseases."
3. "the intentional putting to death by artificial means of persons with incurable or painful disease."
4. "an act . . . in which one person . . . kills another person . . . for the benefit of the second person, who actually does benefit from being killed."[13]

Not only do these definitions differ substantially from one another; they also are open to "fatal counterexamples."[14] Definition 1 could be applied to a simple case of murder, such as injecting a quick-acting poison into a sleeping person whom one hates. This definition also permits accidental death to be an instance of euthanasia.[15] Definition 2 could also be a simple case of murder, such as injecting a quick-acting poison into a rich relative who has a mild case of multiple sclerosis (an incurable disease) without the relative's consent and for the purpose of inheriting his wealth. Definition 3 includes the vague and problematic term "artificial" (what would it mean intentionally to put someone to death by "natural" means?); moreover, like definitions 1 and 2, this definition also omits mention of the central intention, namely, to end suffering or to benefit the person whose death is intended. Definition 4 comes closer to our intuitive understanding of the concept of euthanasia. But cast in terms of one person *killing* a second person, it rules out by definition situations in which suffering patients are passively allowed to die rather than actively killed.

A classification scheme often used to elucidate the concept of euthanasia distinguishes four categories: active, passive, voluntary, and involuntary. Each category can be illustrated by a clear, uncontroversial example. Yet practically speaking, these distinctions are easily blurred. In the following examples, assume that the intention in causing the death of the individual is to end suffering or to benefit the patient.

A clear case of active euthanasia is the injection of a lethal substance into a patient with advanced, metastatic cancer who is suffering extreme pain and discomfort. The act would count as *voluntary*, active euthanasia if the patient knew she was in the advanced stage of incurable cancer, requested the injection, and believed the injection would cause her death. It would be *involuntary*, active euthanasia if the patient was not consulted and her wishes were not known.

Paradigm cases of *passive* euthanasia are ones in which a life-sustaining treatment is withheld. These are cases of allowing death to occur rather than killing. A patient who requests to be allowed to die rather than being intubated and placed on a ventilator might be said to seek voluntary, passive euthanasia. The case becomes more convincing as relief of suffering if the patient had already been on a ventilator, found that mode of existence intolerable, and, after being weaned, requests that she not be reintubated even if it is necessary to prolong her life. To ensure that death by this means would be painless as well as quick, the patient requests that she be adequately sedated.

An example of involuntary, passive euthanasia is withholding renal dialysis from a severely demented or profoundly mentally retarded patient whose kidneys have failed. The act is involuntary because the patient has not requested that dialysis be withheld. It is passive because it is a withholding of treatment. And if the patient is manifestly suffering and has to be tied down and sedated for the thrice-weekly dialysis sessions, each lasting for four hours, the situation would qualify conceptually as euthanasia.

But here is where definitional uncertainty begins, leading to the conceptual debate. What about withdrawing a therapy, such as a respirator, on which a patient is being maintained? Sharp disagreement arises over whether that should count as killing or as allowing the patient to die. Does a "good death" require that the patient be relieved of suffering? Can death itself count as a benefit to a person who is comatose or in a PVS? Reasonable people disagree on the answers to these conceptual questions, yet without agreement, the boundaries of euthanasia remain fuzzy.

According to one view, patients in a coma or PVS "are benefited by euthanasia, even though they are not relieved of suffering."[16] The opposite view holds that "no definition [of euthanasia] is acceptable which includes under its instances persons who are comatose."[17] Supporters of this view claim that the concept of euthanasia should apply only in cases where the intent is to benefit a person by relieving that person's *suffering*. It is important to see this as a conceptual dispute, not an ethical disagreement. There is virtually no controversy over whether individuals who are in a coma or a

PVS are experiencing suffering. Opponents in the conceptual debate might *disagree* about whether withholding life support from comatose individuals counts as euthanasia; yet they may still *agree* on whether it is morally permissible to forgo the respirator.

Conceptual debate is even more intense when it comes to drawing a line between active and passive means. Is the removal of life support already in place active or passive? Many physicians insist that withdrawing treatment counts as active euthanasia. They argue that if they fail to institute life-prolonging treatment, it is the disease that kills the patient, but if life support already in place is withdrawn, it's the doctor who kills the patient. This conceptual maneuver rests on an underlying ethical premise: Allowing the patient to die is morally permissible, but killing the patient is not.

Yet that proposition is problematic. It brings to mind the rationalization used by doctors in Nazi Germany who were involved in Hitler's child euthanasia program. Some of those doctors left their institutions without heat, allowing the children to die of exposure. This enabled the physicians to offer the rationale that they were not engaged in murder, since "withholding care" was simply "letting nature take its course."[18] This shows that ethical permissibility cannot rest simply on whether the means leading to death are active or passive.

At the opposite extreme from supporters of active euthanasia and assisted suicide are those who reject as unethical any and all withholding of life sustaining treatment, whether active or passive, voluntary or involuntary. Then there is a more moderate group who draw the moral line at withholding food and fluids or, more generally, at the boundary between "ordinary" and "extraordinary" treatments or "routine" and "heroic" measures. Withholding food and fluids amounts to starving people to death, it is sometimes argued. Some critics heighten the rhetoric, charging that denying food and fluids even to patients who are dying is the first step down the slippery slope to the practices in Nazi Germany.

The Analogy with the Nazis

No ethical indictment, in medicine or elsewhere, is more devastating than the charge of "Nazi practices." The temptation to make that indictment often surfaces when euthanasia is mentioned. Are there any relevant similarities between "medical killing" during the Holocaust, when Nazi doctors were truly enemies of patients, and current biomedical practices? If there are similarities, does ethical consistency require that we accept or condemn both?

Three general approaches to these questions can be discerned. To be sure, there are differences in style and method among adherents of any one approach. The first approach finds all too many similarities between the Nazi "euthanasia program" and what goes on in hospitals today. The second approach argues against the meaningfulness or the accuracy of alleged similarities. The third approach sounds a cautionary alarm about the dangers of the slippery slope.

An example of the first approach is an inflammatory article entitled "Hitler's Euthanasia Program—More Like Today's Than You Might Imagine." The author begins by asserting: "Today we are faced with the prospect that our society will accept and legalize the crime of euthanasia, for which Nazi doctors hanged at Nuremburg."[19] The article uses the phrase "today's Death Lobby" to refer to advocates of patients' right to refuse life-prolonging treatment and identifies a "whole breed of 'medical ethicists' in the service of death . . . whose vocation is to tell doctors, hospitals, families and clergymen that it is highly moral to kill helpless patients in American hospitals."[20]

Less inflammatory but issuing a similar indictment is Nat Hentoff, who argues that even with the best intentions, it is possible "to think and plan in a way that would bring about results that were also the goals of the Nazis— from different motivations."[21] Hentoff selects several targets. One is Daniel Callahan, for implying in his book, *Setting Limits*, that "the lives of the elderly are worth less—in terms of prolonging them—than other lives," a view Hentoff assimilates to the Nazis' *"lebensunwertes Leben,* 'life unworthy of life.'"[22] Another target is Dr. Ronald Cranford, along with others who argue that when individuals are in a PVS, "personhood" has vanished, and "without personhood, there could be no act of murder" that is, wrongful killing. Cranford was one of the Minnesota physicians who argued in favor of overriding the objections of Helga Wanglie's family to removing her life supports.

A Protestant minister, Richard John Neuhaus, takes the position that not only is euthanasia morally prohibited, but also that it is always wrong to withhold or withdraw certain medical treatments. In an article entitled "The Return of Eugenics," Neuhaus argues that Nazi Germany's doctrines and practices "effected but a momentary pause in the theory and practice of eugenics. . . . [T]oday, . . . eugenics is back."[23] The meaning of "eugenics" is broadened here to include "new ways of using and terminating undesired human life,"[24] as well as other biomedical practices the author deplores. One such practice, abortion, is likened by Neuhaus to the killing of Jews, Gypsies, homosexuals, and Slavs by the Nazis.[25] Here Neuhaus claims that the justifications offered for abortion and research with human embryos in our society "are very much like the arguments employed in the Holocaust."[26]

The second approach to the analogy with the Nazis is at the opposite end of the spectrum. According to this position, there are no real similarities between what the Nazis did in their euthanasia program and examples of forgoing life-sustaining treatment in the United States today. Not only do certain concepts used by the Nazis differ in meaning from their current usage, but also there are great factual dissimilarities between what the Nazis did and what goes on today. One example is a difference in the use of the phrase "quality of life." The Nazis viewed the quality of life of individuals only in terms of their social worth, their service to the *Volk*. In contrast, we view quality of life in terms of individual well-being.[27]

Yet even when its meaning is confined to standards of individual well-being, the phrase "quality of life" poses the problem of determining how high or how low a standard is acceptable. In a long-standing debate surrounding the ethics of allowing handicapped newborns to die, standards have been proposed that many people consider unacceptably high. An example is the criterion used by John Lorber, a British pediatrician: "the ability to work or marry."[28]

A more direct form of this approach focuses on the term "euthanasia" itself. The historian Lucy Dawidowicz claims that such terms, when applied to the Nazi experience, "do not have our meaning. These terms and the programs they stood for were integral aspects of Nazi racism. Nazi racism derived from a theory about the ultimate value of the purity of the *Volk*."[29] Yet Dawidowicz makes only half of the comparison. She fails to specify *our* meaning—assuming that there is a single meaning, which there is not—of the term "euthanasia."

The Slide Down the Slippery Slope

Richard John Neuhaus is one of those who takes the third approach, warning about the dangers of the slippery slope. He contends that even avowed opponents of slippery slope arguments may have unwittingly begun the slide themselves. Neuhaus accuses Daniel Callahan of having already slid:

> Daniel Callahan is a spirited opponent of the slippery-slope metaphor . . . but his own emotional preparedness with respect to the treatment of the dependent and incapable has undergone a remarkable development. . . . In . . . 1983 . . . he wrote forcefully against withdrawing food and water. . . . Four years later, Callahan invites us . . . to discard that moral emotion.[30]

According to Neuhaus, Callahan has slid because to hold that nutrition and hydration may ethically be withheld requires the erasure of a long-standing distinction in medical ethics and practice, the distinction between providing medical treatment and providing food and water.[31] It is easy to see how withholding food and water can be assimilated to Nazi practices. The Nazi doctor Hermann Pfannmuller was credited with the policy of starving to death those selected for the children's euthanasia program rather than wasting medication on them.[32] According to the account of a visitor to the institution Pfannmuller directed, he said:

> "We do not kill . . . with poison, injections, etc.; . . . No, our method is much simpler and more natural, as you see." . . . The murderer explained further then, that sudden withdrawal of food was not employed, rather gradual decrease of the rations. A lady who was also part of the tour asked—her outrage suppressed with difficulty—whether a quicker death with injections, etc., would not at least be more merciful.[33]

The irony of the visitor's question is a reminder of a core ingredient in the meaning of "euthanasia." A lethal injection might very well have brought about a more "merciful" death than starvation for these children. But unlike patients in the last throes of terminal illness, the children in the Nazi euthanasia program were deliberately selected to die. It is a cruel irony to voice an ethical indictment of the *mode* of death when it is the very *fact* of death that constitutes the moral outrage. The same moral judgment applies to the cases of Baby Doe in Bloomington, Indiana, and the Johns Hopkins baby more than 15 years ago. The fact that they were not fed, and hence starved to death, addresses the lesser of the moral evils. Both infants would have lived if their parents had not refused to consent to surgery to correct their intestinal malformation. And both would have lived had a court order been granted to override parental refusal. There is every reason to believe that these infants suffered.

But there is no reason to believe that irreversibly comatose patients or those in a PVS suffer when artificial nutrition and hydration are withheld. These patients have no awareness of their surroundings, nor do they experience sensations. They do not respond to any stimuli, even deep pain. The part of their brain that enables them to have experiences has ceased to function. Therefore, they cannot suffer, either from starvation, dehydration, or continued medical treatment. Whether withdrawal of life supports from such patients should be termed euthanasia is a conceptual question. Whether withdrawal of medical treatment, including artificially administered nutrition and hydration, is ethically acceptable is a separate and distinct question.

One ethicist observes that there's more than one slope to slide down. Laurence McCullogh argues that the accuracy of the comparison between the Nazi euthanasia program and our own biomedical practices depends on which features are taken to be relevantly similar or different:

We see that what got German society on the slippery slope, indeed what characterized the slope, was the racist attitudes already in place. It is a reasonable defence
and distinguishes our case from theirs, for us to say that we don't have those
attitudes. . . . *Our* slippery slope might yet be analogous to Germany's in a more
abstract way. If we consider the rationale which gives social utility or economic
returns precedence over individual freedom, then we might see how our society
could approach the kind of thinking that underlay the Nazi experience. There,
racism overrode personal autonomy; here, it might be an economic rationale—the
attitude that we won't spend so much per year to keep somebody alive on the slim
chance of recovery.[34]

This view is echoed and reinforced by Alan Weisbard and Mark Siegler,
who observe that one slide has already begun, and now is the time to be
vigilant. Their concern is the acceptance by the medical profession and by
many biomedical ethicists of withholding food and fluids from dying patients, conjoined with recent furious efforts to control the costs of medical
care:

We have witnessed too much history to disregard how easily society disvalues the
lives of the "unproductive"—the retarded, the disabled, the senile, the institutionalized, the elderly—of those who in another time and place were referred to as
"useless eaters." The confluence of the emerging stream of medical and ethical
opinion favoring legitimation of withholding fluids and nutrition with the torrent of
public and governmental concern over the costs of medical care . . . powerfully
reinforces our discomfort.[35]

Concern about still another slippery slope is sometimes voiced regarding
the role physicians are asked to play in voluntary, active euthanasia. If
doctors begin to perceive an obligation to kill patients who request it, in
addition to their long-standing obligation to heal their patients, they could
readily develop an indifference to human beings. Robert Jay Lifton identifies
a psychological process he calls "doubling," whereby Nazi doctors could
function in their evil role. Lifton describes the process as follows:

the division of the self into two functioning wholes, so that a part-self acts as an
entire self. An Auschwitz doctor could, through doubling, not only kill and contribute to killing but organize silently, on behalf of that evil project, an entire self-
structure (or self-process) encompassing virtually all aspects of his behavior.[36]

Whether doubling, or becoming indifferent to the value of human life, is an
inevitable consequence of physicians performing active euthanasia in response to patients' requests is a factual question, one that can only be answered by observation and experience.

Similarly, evidence is needed for the assertion that a slide down a parallel
slope is imminent: that physicians will begin by performing voluntary eu-

thanasia and then readily shift to engaging in involuntary euthanasia. Indeed, critics of the practice in the Netherlands claim that already there are numerous instances of involuntary euthanasia and predict that more will follow, despite the strict safeguards designed to permit only voluntary euthanasia with terminally ill patients.

Some find the possibility of allowing active euthanasia—even the voluntary kind—to be the most worrisome prospect. Others fear that from an ethical standpoint, the most dangerous practice is withdrawing or withholding treatment involuntarily, that is, from patients who lack decisional capacity. I maintain that a far greater danger of a slide down the slippery slope lies in the use of an economic rationale for deciding when to withhold or withdraw treatments from patients.

When the justification offered for terminating treatment is that it is not "costworthy," or that it is consuming a disproportionate amount of society's or an institution's resources, the slide down one of the slopes to the Nazi program has already started. As one scholar has observed, "the argument for the destruction of life not worth living was at root an economic one."[37]

This is not to overlook the salient fact that racial ideology was a driving force behind Nazi policies. Is our society so free of racism that we are in no danger of sliding down that slope as well? I think that danger may lurk in the background, but it is likely to arise in an indirect fashion rather than directly. Although racial prejudice in our society undeniably exists, it does not assume the proportions of an ideology except in the case of fringe sects and fanatical groups. However, since disproportionate numbers of the poor are members of racial or ethnic minorities, a de facto pattern of discrimination could occur when economic factors influence decisions to terminate treatment.

Euthanasia and Public Policy

Actual legislative proposals have already appeared in this country, one several years ago in California and a more recent one in Washington State. A petition sponsored by a group called Americans Against Human Suffering sought to place a referendum legalizing active euthanasia on the California ballot in 1988. The group failed to gather the 450,000 signatures needed to place the referendum on the ballot, despite the fact that a poll taken in California in 1987 revealed that 64 percent of the state's residents favored active euthanasia for the terminally ill. Initiative 119 appeared on the Washington State ballot in November 1991. Although surveys had shown sufficient support among voters to approve the bill, it was defeated in the November election. Derek Humphry announced that the Hemlock Society was already gathering signatures for initiatives that would appear on the ballot in 1992 in California and Oregon.[38] All legislation proposed so far permits only *voluntary* euthanasia. Additional safeguards, designed to prevent abuses, are built into the proposed legislation.

I believe that the action of Dr. Timothy Quill in assisting his patient, Diane, to commit suicide was ethically acceptable, courageous, and even commendable. I indicated earlier the ways in which Dr. Quill's assisting his patient's suicide differed from Dr. Kevorkian's unethical action in helping Janet Adkins commit suicide. Therefore, it cannot be the mere fact that a physician provides aid in dying to a patient that makes some actions of this type wrong. What made Dr. Kevorkian's actions wrong were his brief and scant knowledge of Janet Adkins, the absence of independent evaluation of her diagnosis and mental state by at least one other physician, and the absence of manifest suffering or imminent physical deterioration of the patient.

Despite my judgment that Dr. Quill's action and others like it are ethically acceptable, I remain opposed to enacting laws that would authorize physicians to perform active euthanasia. (In referring to "active euthanasia," I do not mean to include a physician writing a prescription for barbiturates or similar drugs, as Dr. Quill did for his patient.)

My opposition to legalizing euthanasia is open to criticism on at least two counts. First, it might be considered inconsistent because it fails to support patients' right to self-determination, a right I have been championing all along. And second, opposing legalization could further the gap between the treatment of patients who have a personal physician of long standing and that of patients whose medical care is delivered in hospital outpatient clinics or emergency rooms, and who thus lack access to a doctor who knows them and will advocate for their interests.

Is it ethically inconsistent to commend the actions of Dr. Quill and other physicians who assist their patients' voluntary request for assisted suicide or active euthanasia and, at the same time, oppose the enactment of laws that would legalize such actions generally? I think not, for several reasons. First and foremost, it should remain difficult, not become easy, for physicians to actively assist their patients to die. The medical profession's traditional respect for life and the training of doctors to strive to preserve life should not be faulted because modern technology allows physicians to prolong unwanted life.

Efforts should always be made to eliminate patients' physical pain. In an elderly patient or one whose respiratory system is compromised, a high dose of narcotics also has the consequence of suppressing respiration and could possibly hasten death. That should not be a reason for withholding pain medication from such patients. And doctors should certainly respect their patients' right to refuse burdensome or unwanted medical treatment. If conscientious efforts are made, there should be only a small number of patients whose pain and suffering is so manifest that they request the active assistance of physicians in terminating their lives.

But that number cannot be reduced to zero. For some people, the choice between being in constant pain or being drugged or asleep as a result of medication is unacceptable. To remain alive but unconscious is not a genuine choice, since it is the virtual equivalent of death. For others, total paralysis with no ability to function and no hope of recovery is an intolerable fate. For physicians to provide aid in dying to these patients is ethically acceptable, perhaps even desirable, and should be possible without a likelihood of their being indicted and convicted of homicide. It should be possible without having to authorize a form of institutionalized killing by physicians.

The current state of affairs seems to work reasonably well in most respects. Dr. Kevorkian was charged with murder, but a judge in the Oakland (Michigan) County Court dismissed the charges against him. The Oakland County prosecutor then decided to drop the murder case against the physician, saying that euthanasia and suicide are issues for the state legislature, not the police and the courts. Nevertheless, the prosecutor criticized the judge for dismissing the charges. For his part, the judge ruled that Dr. Kevorkian should not stand trial because there was no proof that he had planned and carried out the death of Janet Adkins.[30] Some months later, in a civil case over the use of the suicide machine, a different circuit court judge banned Dr. Kevorkian from using his device again. The judge contended that Dr. Kevorkian's "goal is self-service rather than patient service. . . . [He] has a propensity for media exposure and seeks recognition through bizarre behavior."[40]

In the case of Timothy Quill, a grand jury failed to indict the physician for murder. Traditionally, juries have acquitted physicians whose actions were unmistakably mercy killings. Another example was that of Dr. Vincent Montemareno, a New York physician who in 1974 was charged with giving a lethal injection of potassium chloride to a patient with terminal cancer. The jury voted to acquit him.[41] Also, as doctors themselves acknowledge, cases regularly occur in which physicians help their patients die, cases that never come to public attention because they are carried out in the privacy of a patient's home or within the confines of a well-established doctor–patient relationship.

What about the argument that a form of injustice results from continuing to prohibit assisted suicide? Patients likely to avail themselves of a physician's assistance when active euthanasia is illegal are people like Janet Adkins, educated and having the financial means to travel to Michigan to obtain assistance from Dr. Kevorkian; and like Dr. Quill's patient, Diane, who had a personal physician who cared for her for eight years and who also knew her family. Education and social class distinguish these patients from the many people who lack a personal physician, who obtain their medical care in emergency rooms or outpatient clinics, or who do without regular medical care altogether until they are hospitalized with a life-threatening condition. Isn't it an unjust state of affairs when financially well-off and better-educated patients are much more likely to obtain physician-assisted suicide than those who are poorer or less well educated?

Yes. But the likely consequences of rectifying this injustice are probably worse than the injustice itself. If there is genuine cause for concern about possible abuses of a law legalizing active euthanasia, this concern is heightened in the case of vulnerable populations and disenfranchised citizens. Moreover, anecdotal evidence and informal reports from a hospital that developed a living will for AIDS patients revealed that poorer patients, who generally acquired the disease through intravenous drug use or by having sex with drug-using partners, were uninterested in signing documents that would authorize termination of treatment. Having had less access to medical services for most of their lives, these patients were more likely to want everything to be done than to insist on termination of life-prolonging treatments so that they could experience death with dignity.

An array of specific, highly undesirable consequences of euthanasia legislation has been suggested.[42] For the most part, these are not actual, observed consequences but conjectured ones. Two overriding concerns that have been expressed are (1) likelihood of mistake and abuse and (2) the danger that legal machinery initially designed to kill those who are a nuisance to themselves may some day engulf those who are a nuisance to others.[43]

What about the possibility of some sort of error: a faulty diagnosis, a mistaken prognosis, or the possibility that relief or even a cure may be found, especially when the patient does not have a rapidly deteriorating condition?[44] Proponents of legalizing euthanasia offer replies to all these objections, contending either that erecting safeguards can succeed in preventing these bad consequences or that the predictions of their likelihood have little or no basis.

A different type of worry focuses on the voluntariness of the patient's decision. Might not some people be persuaded to request euthanasia? Patients might be influenced by their relatives to avail themselves of the legal alternative of euthanasia. Or even if relatives do not actually try to persuade a gravely ill patient to opt for euthanasia, the patient may decide that those relatives would be better off. Some may come to believe that they owe it to their families to hasten death and relieve their relatives of financial or emotional burdens. If the expressed wish for euthanasia is coerced, even by the most subtle means, it is not an ethically sound option.

Another concern has been voiced about whether patients who are heavily drugged by narcotics are in a proper mental state to request euthanasia. In the alternative case, before heavy doses of narcotics have been administered, patients racked with pain are hardly in a position to think clearly about life-and-death choices.[45] This concern addresses the rationality of the decision, contending that the patient is either in too much pain or too drugged to make a rational decision about death. Although it is probably impossible for anyone to make a rational decision once a disease has progressed to that point, a patient could very well begin to prepare for that eventually at a much earlier stage, after the fatal illness has been diagnosed but before the inability to think rationally has set in.

To institutionalize the practice by laws that authorize active euthanasia would, I believe, open the door to a number of dangers. We saw earlier how physicians have managed to circumvent various provisions of a state law on DNR orders by finding creative ways to do what they think proper, regardless of a patient's or family's wishes, by perpetuating deceptive practices such as show codes and by locating another doctor who will document something in a chart. Laws are always open to violation or abuse, but a law that changes the presumption from "doctors must not kill" to "doctors may kill under the following conditions" appears too dangerous to endorse with any comfort.

Some commentators have remarked that it is not good public policy to have a gap between laws prohibiting euthanasia and what happens when alleged violators are apprehended. For example, the philosopher James Rachels argues that it is undesirable for such a wide gap to exist: "If we deem a type of behavior not fit for punishment, why should we continue to stigmatize that behavior as criminal?"[46] Rachels says the most common answer is that "by maintaining the legal prohibition, we safeguard the important principle of respect for life that would otherwise be eroded."[47] But that answer is somewhat of a platitude and requires further elaboration. I am among those who think that tolerating the gap between law and common practice is better than the alternatives: either legalizing active euthanasia or indicting and convicting physicians of murder.

Derek Humphry argues that failure to enact laws legitimating active euthanasia guarantees that doctors will be enemies of their dying patients, or at least that they cannot be proper advocates. He believes that if they contain proper safeguards, such laws would pose no dangers from well-meaning physicians. Evil doctors, like evil people everywhere, will do their malevolent deeds whatever the law may enjoin. But good doctors should be enabled to act compassionately on behalf of their patients without fear of criminal liability.

Advocates of active euthanasia contend that doctors who fail to assist their patients in dying are seen by patients as the enemy. Precisely the opposite judgment was made by a physician in Sweden who refused his patient's request to inject potassium into his veins. Using a common and possibly sound objection to doctors' performing active euthanasia, he cited the "physician's relationship with his patients. If physicians start to kill, then there may be no end. Everyone would be afraid and distrustful."[48] This view warns that if doctors as a rule begin to engage in active euthanasia, the entire medical profession would come to be seen as the enemy. Clearly, more evidence needs to be gathered about people's actual beliefs regarding the physician's role in assisted suicide and euthanasia.

There is surely nothing to fear from good and compassionate doctors, whose sole motive is to act in the best interest of their patients. But few of us ever act solely from a single motive, especially a selfless or noble one. Medicine is a powerful profession, and physicians have a long tradition of exercising their authority and commanding respect for their expertise. I have no doubt that all of the physicians who sought to remove Helga Wanglie's ventilator were good and compassionate. But recall that they were not seeking to remove Mrs. Wanglie's respirator for *her* sake, but rather for their own sake as medical professionals: Keeping patients in her condition alive is not good medical practice. Similarly, the many physicians joining together to assert their decision-making authority in cases they deem medically futile do not have base motives. But neither are their motives patient centered.

When we think back on the entire picture—physicians who write DNI orders when they are prevented from writing DNR orders; physicians who refuse to replace heart valves in HIV-positive drug addicts, physicians who are pressured by hospital administrators or intimidated by risk managers into acting against the interests of their patients; physicians who deny a patient an intensive care unit bed because a better candidate might come along later that night—it is not a picture that portrays doctors as always acting as unswerving advocates for their patients.

It is ethically sound to commend those physicians who act on an individual, compassionate basis in assisting their patients to end their lives. At the same time, it is ethically defensible to oppose legally sanctioning such behavior as a matter of public policy. The result is that the right of some—perhaps many—people to self-determination of their "final exit" may be denied them. My argument against legalizing active euthanasia is based on the potentially bad consequences that could very well result from having that public policy.

Admittedly, this is a version of the slippery slope argument, a form of reasoning that is sometimes sound and at other times not. Its soundness depends on the factual premises: the accuracy of predictions about whether a slide down an undesirable slope is likely or inevitable, and just what will happen if the slide does take place. The undesirable consequences of an irreversible slide down the wrong slope seem to me worse than the bad consequences of failing to accord suffering or dying patients the legal right to assistance by a physician in ending their life.

Anyone who tentatively accepts the ethical permissibility of voluntary, active euthanasia would almost certainly reject its incorporation into social policy if it became evident that any of the negative predictions about the slippery slope were likely to be realized. Some actions might be morally right in an ideal world but should be ruled out in the real world, given human fallibility and other practical deterrents to attaining the ideal.

Notes

1. Nancy W. Dickey, "When the Doctor Gives a Deadly Dose," Commentary, *Hastings Center Report*, vol. 17 (1987), p. 34.

2. Leon R. Kass, "Neither for Love Nor Money: Why Doctors Must Not Kill," *The Public Interest*, Number 94 (Winter 1989), pp. 36–37.

3. Ibid., pp. 37–38.

4. "Doctor's Suicide Device Used," *The Los Angeles Daily Journal*, Section II (June 8, 1990), p. 1.

5. "A Conversation with Dr. Jack Kevorkian," *Ann Arbor News* (June 24, 1990), p. C11.

6. Howard Brody, M.D., as quoted in Elisabeth Rosenthal, "In Matters of Life and Death, The Dying Take Control," *New York Times*, Section 4 (August 18, 1991), p. 1. (Emphasis added.)

7. Timothy E. Quill, "Death and Dignity," *New England Journal of Medicine*, vol. 324 (1991), pp. 691–94.

8. Ibid., p. 692.

9. Ibid., p. 693.

10. Ibid.

11. Ibid., p. 694.

12. The following discussion of the concept of euthanasia and the Nazi euthanasia program is adapted from my essay "Which Way Down the Slippery Slope: Nazi Medical Killing and Euthanasia Today," in Arthur Caplan (ed.), *When Medicine Went Mad* (Totowa, N.J.: Humana Press, 1992), pp. 173–200.

13. Tom L. Beauchamp and Arnold I. Davidson, "The Definition of Euthanasia," reprinted in Samuel Gorovitz, Ruth Macklin, Andrew L. Jameton, John M. O'Connor, and Susan Sherwin (eds.), *Moral Problems in Medicine*, 2nd ed. (Englewood Cliffs, N.J.: Prentice-Hall, 1983), pp. 446, 447, 457.

14. Ibid., pp. 446ff.

15. Ibid., p. 449.

16. Ibid., p. 453.

17. Ibid.

18. Robert N. Proctor, *Racial Hygiene: Medicine Under the Nazis* (Cambridge, Mass.: Harvard University Press, 1988), p. 187.

19. Molly Hammett Kronberg, in Nancy B. Spannaus, Molly H. Kronberg, and Linda Everett (eds.), "How to Stop the Resurgence of Nazi Euthanasia Today," *EIR Special Report* (September 1988), p. 129.

20. Ibid., pp. 130–31.

21. Nat Hentoff, "Contested Terrain: The Nazi Analogy in Bioethics," *Hastings Center Report*, vol. 18 (1988), p. 29.

22. Ibid.

23. Richard John Neuhaus, "The Return of Eugenics," *Commentary*, vol. 85, No. 4 (1988), p. 15.

24. Ibid.

25. Richard John Neuhaus, "The Way They Were, the Way We Are," in Caplan (ed.), *When Medicine Went Mad*, pp. 211–30.

26. Ibid., p. 228.

27. Cynthia B. Cohen, "'Quality of Life' and the Analogy with the Nazis," *The Journal of Medicine and Philosophy*, vol. 8 (1983), p. 114. See also Cohen's commentary in "Contested Terrain," pp. 32–33.

28. John Lorber, "Ethical Problems in the Management of Myelomeningocele and Hydrocephalus," *Journal of the Royal College of Physicians*, vol. 10 (1975), pp. 47–60.

29. Lucy Dawidowicz, as cited in Helga Kuhse and Peter Singer, *Should the Baby Live?* (New York: Oxford University Press, 1985), p. 94.

30. Neuhaus, "The Return of Eugenics," p. 24.

31. Ibid., p. 22.

32. Robert Jay Lifton, *The Nazi Doctors: Medical Killing and the Psychology of Genocide* (New York: Basic Books, 1986), p. 62.

33. Ibid.

34. Laurence McCullogh, cited in David Lamb, *Down the Slippery Slope: Arguing in Applied Ethics* (New York: Croom Helm, 1988), p. 29.

35. Alan J. Weisbard and Mark Siegler, "On Killing Patients with Kindness: An Appeal for Caution," in John D. Arras and Nancy K. Rhoden (eds.), *Ethical Issues in Modern Medicine*, 3rd ed. (Mountain View, Calif.: Mayfield, 1989), p. 218.

36. Lifton, *Nazi Doctors*, p. 418.

37. Proctor, *Racial Hygiene*, p. 183.

38. Jane Gross, "Voters Turn Down Mercy Killing Idea," *New York Times* (November 7, 1991), p. B16.

39. William E. Schmidt, "Prosecutors Drop Criminal Case Against Doctor Involved in Suicide," *New York Times* (December 15, 1990), p. A10.

40. Associated Press, "Kevorkian Permanently Barred from Using His Suicide machine," *Ann Arbor News* (February 5, 1991), p. A1.

41. Case cited in James Rachels, *The End of Life: Euthanasia and Morality* (New York: Oxford University Press, 1986), p. 170.

42. The consequences noted here are all taken from Yale Kamisar, "Euthanasia Legislation: Some Non-Religious Objections," reprinted in Gorovitz et al., *Moral Problems in Medicine*, pp. 458–63.

43. Ibid., p. 459.

44. Ibid., p. 461.

45. Ibid.

46. James Rachels, *The End of Life: Euthanasia and Morality* (New York: Oxford University Press, 1986), p. 170.

47. Ibid.

48. Carl M. Kjellstrand, "The Impossible Choice," *Journal of the American Medical Association*, vol. 257 (1987), p. 233.

10

Ethics Committees as Advocates for Patients

Is there anyone left to serve as proper advocates of patients? Probably only a few people are aware that in the United States today, most hospitals have established committees to deal with ethical issues. Some of these committees tackle individual cases involving patients, making recommendations that lead to a future decision or reviewing past cases that caused ethical controversy. Other committees concentrate on preparing draft policies or developing guidelines for ongoing ethical concerns that could benefit from a systematic mechanism. At least one state (Maryland) has enacted a law requiring all hospitals to have "patient care committees," and laws in other states, such as New York, mention ethics committees as a desired mechanism for resolving conflicts.

What do ethics committees do? And what should they be doing? I believe that the purpose of having ethics committees in hospitals is to help ensure that the rights of patients are protected and that their interests are promoted. If that purpose is not accomplished, then ethics committees are wasting a lot of busy people's valuable time. But if that purpose is successfully served, others besides patients can benefit from the committee's work. By providing a systematic framework for ethical decision making and by engaging in thorough and reflective deliberations, committees help to educate the clinicians who bring cases for review. Those committees that have members who are well acquainted with the literature in bioethics and health law can even enlighten hospital administrators.

When ethics committees become involved in individual cases in the hospital, some physicians object, claiming that ethics committees intrude on the physician–patient relationship just as much as the other uninvited intruders discussed earlier. These committees are supposed to serve as advocates of patients' rights and interests. Yet intruders can be advocates, whether they are invited or imposed. When judges become involved in cases, they intrude into the physician–patient relationship. But by intruding, they can protect or even establish a patient's rights, as they have in cases involving the right to refuse medical treatment. Similarly, when ethicists or ethics committees act as intruders into the physician–patient relationship, they can still play an advocacy role. Not all intruders should be viewed as enemies of patients.

Some Examples: A Look Back

Earlier chapters portrayed a number of different situations in which ethics committees were involved in cases. A look back at these examples reveals that the actual workings of committees can vary considerably. Sometimes the committee has a strong voice and can exert influence. At other times the committee may draw a conclusion about the ethically right thing to do, but it is powerless to affect the ultimate decision. And at still other times, the ethics committee is divided and can reach no substantive conclusion leading to a recommendation.

In the case of Marjorie Brown, the patient with Lou Gehrig's disease who requested removal of her respirator, the committee deliberated for many months. A small subcommittee engaged in more than one bedside consultation; the entire committee reviewed different aspects of the case; and there were ongoing communications with the hospital's risk manager and lawyers from the corporate governing board. Discussions focused on the complex relationship among the various parties involved: the ethics committee itself, the hospital administration, the corporate governing body, the clinicians involved in managing the patient's care, and, of course, the patient. Despite authorization from the corporate board, the patient's request still had not been honored several months after the case had been brought to the committee because of unexplained delays by the administration.

Eventually, at one of the regular monthly meetings, some committee members became irate. One person asked: What is the appropriate role of the hospital ethics committee in monitoring the hospital administration's facilitation and the clinicians' implementation of the patient's wishes? Given the fact that five months had elapsed from the time the committee had first become involved, should the committee assume a more active role of advocacy for the patient? Although hospital ethics committees should normally not stand in an adversary relationship to the hospital administration, the committee's role in protecting the rights and promoting the interests of this patient forced it into precisely this position.

In the similar case of the dying AIDS patient who asked that his respirator be removed and that he be given sufficient pain medication to prevent suffering, the hospital administration became involved at a very early point. The hospital's CEO was the one who eventually decided what should happen. He prevented the physician from administering morphine to sedate the patient before removing the respirator, insisting that morphine could be given only when the patient's suffering became manifest. In both the case of Marjorie Brown and that of the AIDS patient, hospital ethics committees were involved before the final decision was made, but with no ultimate authority to act. The involvement of the committee in the second case was a mere formality, since the risk manager and in-house counsel were brought in at an early stage, and the hospital's CEO was alerted soon thereafter.

In the Linares case in Chicago, in which the hospital steadfastly refused to remove the comatose infant from the respirator, the issue was never brought to the ethics committee. Max Brown, the hospital's lawyer, said he saw no point in doing so. He said: "What we're dealing with here is a legal problem. Ethics committees are fine so long as what is ethically being contemplated is legally acceptable. When you have an ethical alternative that is by all accounts illegal, an ethics committee cannot do much to make it legal."[1]

However, others familiar with the situation point out that although the hospital did have an ethics committee, it was a nonfunctioning committee. The Linares family was never told that an ethics committee existed or that they might have access to it. If the committee had been properly constituted and well functioning, the Linares case might very well have turned out differently. The presence of an academic attorney or a bioethicist would have compelled Max Brown to take a closer look at the laws he so blithely invoked.

Brown expressed absolute certainty about what the law did and did not allow, a certainty not shared by other lawyers who have looked at Illinois law. The judgment that a particular course of action is prohibited by law could be the final conclusion in a problematic case. But there are good reasons for bringing the case to an ethics committee for review and discussion at an earlier point, especially since someone on the committee might have a different and knowledgeable opinion about the state of the law. Beyond that, it is worthwhile to review cases from a purer ethical standpoint. Sometimes laws are changed for ethical reasons, and those reasons might very well be raised in the deliberations of a thoughtful and informed ethics committee.

Then there was the case of the intravenous drug user who had reinfected his heart valve and whose surgeons were refusing to operate a second time. This was not the first occasion on which cases of this type were brought to our hospital ethics committees. In fact, this same surgeon had brought a similar case to the meeting of a committee at another affiliated hospital. He decided to bring these cases to the ethics committee for guidance, and to see whether the committee shared the position now apparently held by a majority of surgeons.

This was not a prospective review designed to lead to a recommendation in the particular case. It was, rather, an ongoing ethical issue about which the surgeon thought the committee might be able to enlighten him. But committee members were divided. They were divided for all the reasons that might be expected: because the issue of drug use and the behavior of addicts involve deep resentment and social prejudices; because many physicians sympathize with the plight of their surgical colleagues, especially the fact that there is some risk of their becoming HIV infected; because other features that add up to a poor prognosis for the patient raise the question of whether taking that risk of becoming infected is warranted; because committee members are persuaded that all hospital resources should be considered scarce since they are expensive; and because replacing the heart valves of drug addicts isn't worth the cost to the hospital or to society.

The surgeon concluded by asking: What about the patient who is found taking drugs through an intravenous line while still in the hospital? Should that patient be given a repeat operation if the valve becomes reinfected? Most ethics committee members agreed that this would be the limiting case, but their conclusion seemed to be based on their feelings about drug addicts and the antisocial behavior of addicts both inside and outside of hospitals, rather than on a principled moral argument.

Another case described earlier was that of the 70-year-old patient who had been brought to the emergency room with a temperature of 107°F. Following her admission to the hospital seven months earlier, she was given a diagnosis of status epilepticus. The physician who brought the case to the ethics committee was being pressured by hospital administration to decrease the level of care she was receiving. He felt uncomfortable because the family insisted that the care the patient was receiving be continued, and the doctor felt obligated to go along with the family's request.

The physician told the ethics committee that the patient and her husband were Holocaust survivors, having spent years in a concentration camp. The family's experience under the Nazis underscored their unwillingness to forego life-prolonging treatments. Some members of the ethics committee wondered whether it was ethically relevant that these people were Holocaust survivors. Physicians at the meeting were in unanimous agreement that the patient's death would be hastened if she received anything less than her current level of care. To be sure, the intention of the hospital administration in seeking to lower the level of care was not to cause the death of the patient but to save money.

One member of the ethics committee suggested that the family be told that the hospital simply did not have the resources to continue to provide this level of care for the patient. Everyone agreed that continued biological life could not benefit the patient herself. Some committee members concurred with the judgment that this was a waste of the hospital's resources and that such demands by a patient's family should not be honored. But other committee members were deeply worried about the justification for withdrawing treatment. This patient was costing the hospital too much money. Like the psychiatric patients and the retarded children in Nazi Germany, she had become a useless eater.

What did the fact that the patient was a Holocaust survivor have to do with the ethics of lowering her level of care? True, it was a poignant irony to witness a hospital administration offer the same rationale for ceasing to provide treatment to a Holocaust survivor that was used in Nazi Germany to withdraw medications, food, and other necessities of life from mentally disabled Holocaust victims. Yet the fact that the patient and her husband were survivors of the Holocaust had nothing to do with the argument I made at the committee meeting in support of the physician's obligation to the family. My argument was that the practice of rationing medical care at the bedside is morally flawed because it leads to great injustice and will not succeed in shifting monetary resources to other patients in need. Opponents argued that families should not be permitted to demand "unreasonable" care for their incapacitated relatives.

Although this case was not presented to the committee as one that required a resolution and a subsequent recommendation, it was evident from the remarks of those present that no clear consensus was likely to emerge. The only role the committee actually played was that of providing a forum for debating the issues and perhaps providing ethical comfort to the physician. Had the committee sought to develop a policy that would govern similar cases in the future, it probably would have failed to reach agreement.

The Policymaking Role of Ethics Committees

Many ethics committees have adopted as a primary function (instead of or in addition to reviewing cases) that of proposing policies or preparing guidelines for ethical problems that arise in the hospital. Examples of policies devised by hospital ethics committees include those pertaining to DNR orders, guidelines for the termination of life support, policies relating to the care and treatment of AIDS patients, and policies concerning Jehovah's Witnesses who refuse blood transfusions. As in its function of reviewing cases, a committee's policy function should be directed at ensuring that the rights of patients are respected and their best interests served. However, because no policy devised by an ethics committee will ever be adopted without the approval and official sanction of the hospital administration, well-intended efforts by the committee may be sabotaged by those in power, based on risk management concerns or other administrative objections.

One of the hospital ethics committees on which I sit became engaged in a lengthy and difficult process, beginning with a request by a hospital administrator.[2] A problem had arisen with a Jehovah's Witness patient, an emancipated minor, who was admitted to the hospital and refused any transfusions that might be judged necessary in the course of surgery. The patient was alert and lucid, and since the surgery was not an emergency, there was time to negotiate consent. Present in the hospital were the patient's parents, both Jehovah's Witnesses, who were seeking to influence their son not to consent to a transfusion even if needed to save his life. The attending physician was uncertain how to proceed. Aware that judges routinely override parental refusal to consent to blood transfusions for minor children, the physician wondered whether it was appropriate to seek a court order. But noting that this patient was an emancipated minor, the doctor recognized the patient's legal standing to decide for himself.

The attending physician contacted the hospital's risk manager for advice on how to proceed. The case was complicated further by the anesthesiologist's unwillingness to administer anesthesia under these circumstances and by the involvement of the director of the hospital's blood bank, who has the responsibility to supply blood for patients about to undergo surgery While these several concerned parties were deliberating about what to do, the attending physician seized an opportunity to have a thorough discussion with the patient when his parents were not present. The patient agreed to sign a consent form to receive blood. In what was clearly the best outcome, the surgery was performed without the need for a blood transfusion. Following this trying episode, the hospital's risk manager requested that the ethics committee review the existing guidelines and, if necessary, draft a new hospital policy.

The task was to come up with a policy dealing with refusals by Jehovah's Witnesses of blood transfusions, a policy that would recognize the right of adult patients with decisional capacity to refuse medical treatment and, at the same time, protect the interests of children and incapacitated patients from decisions by relatives that could result in harm to them or even death. The committee began by educating itself about the religious, clinical, and legal issues surrounding Jehovah's Witnesses' refusals of blood transfusions.

I informed the committee that their prohibition derives from biblical texts. The New Testament (Acts 15:19–21) restates the Old Testament's prohibition against eating blood or flesh with blood in it:

> And whatsoever man there be among you, that eateth any manner of blood: I will even set my face against that soul that eateth blood, and will cut him off from among his people. (Leviticus 17:10–14).

In their widely quoted reference, *Blood, Medicine, and the Law of God*, Witnesses constantly refer to the medical printed matter of the early twentieth century, which declared that blood transfusions are nothing more than a source of nutrition by a shorter route than ordinary. The consequences of violating the prohibition are dire: Receiving blood transfusion is a sin, resulting in withdrawal of the opportunity to attain eternal life.

Physicians at the meeting commented on some of their concerns. Patients who are ambivalent about refusing a recommended transfusion may, under the stress of being in the hospital and about to go to surgery, decide to accept a transfusion. Some even indicate their wish to be coerced, so that the transfusion is administered, in words often uttered by Jehovah's Witness patients, "against my will." Conversely, some patients strongly desire the transfusion but, in the presence of a church elder or under pressure from a family member, continue to refuse despite their wish to live. The committee decided that the new policy should address these concerns.

An attorney-ethicist who serves on the committee provided some legal background. In general, courts have not identified Jehovah's Witnesses' refusal of lifesaving blood transfusions as manifesting suicidal intent. Most court decisions have permitted physicians to transfer such patients to other physicians and thereby avoid a charge of abandonment of the patient. Courts have not permitted congregation members to refuse transfusions for their minor children, as that would be tantamount to medical neglect. A fact was brought out that was to become central to the committee's work on this policy: There have been very few reported legal cases on the ability of pregnant Jehovah's Witnesses to refuse care.

Without much difficulty, the committee came to agree on details of the policy relating to the vast majority of cases: adult Witnesses with capacity; those lacking capacity who had clearly and unequivocally stated their wishes about transfusions prior to losing capacity and those who had never stated any prior wishes; infants and children; and adolescents near the age of majority. The committee drew upon fundamental ethical principles in formulating the policy for these classes of patients, and added procedural safeguards for patients who appeared ambivalent or whose families seemed to be exerting pressure on them to refuse blood and for emergency situations. The drafting committee presented a first draft of the policy at a monthly meeting of the whole committee.

The drafting committee reported an especially challenging feature of its task: While remaining committed to protecting the right of competent adult patients to reject lifesaving treatment, we believed that a very high standard is required to establish the existence of a Witness's firmly held religious conviction, given the potentially catastrophic and irreversible consequences of refusal of blood. Three situations were of particular concern: where coercion from third parties could influence the decision of patients who had the capacity to decide for themselves; where incapacitated patients are subject to decision making by coreligionists in the family; and where the patient is a pregnant woman.

The first version of the policy drawn up by the drafting committee was discussed in detail. The section on children was scrutinized carefully. The original draft read as follows:

> Parents have the right to consent to care for their dependent children; they do *not* have a coequal right to refuse care for their dependent children (Family Court Act. Section 233). Parents do not have the right to deny minor children transfusions that are deemed medically necessary. In the event that a parent withholds consent for transfusion, the hospital administration must be contacted immediately and asked to seek a court order for transfusion. In the event that medical judgment holds any delay to be immediately life-threatening to the child, transfusion should be given.

A pediatrician on the committee observed that proper advocacy for infants and children demanded that the policy cover the possibility of serious harm, as well as life-threatening circumstances. Others endorsed that view, so the last sentence of the section on children was changed to read:

> In the event that medical judgment holds any delay to be immediately life-threatening to the child or *would produce irreversible harm*, the transfusion should be given.

At the next monthly meeting, the seeds of controversy about pregnant women began to erupt. The original draft of the policy had adopted the framework set out in the U.S. Supreme Court's decision in *Roe v. Wade:*

> In the case of a pregnant Jehovah's Witness's refusal of transfusion, policies relating to adult patients will apply before the third trimester. If the pregnancy has entered the third trimester, the State's interest in life requires the decision to be referred to the courts for adjudication. If in such a case medical judgment holds any delay to constitute an immediate threat to the pregnancy, transfusion should be given.

After some deliberation, the committee decided to alter the section on pregnancy for a reason that had more to do with legal language than with the rights of pregnant patients. In order to conform to judicial wording, a revision was proposed that abandoned talk of "trimesters" and referred instead to the number of weeks pregnancy had advanced. The newly worded section read as follows:

> In the case of a pregnant Jehovah's Witness's refusal of transfusion, policies relating to adult patients will apply before the twenty-fifth week. Beyond the twenty-fifth week the decision should be referred to the courts for adjudication.

My own position at that time was to accept, naively and uncritically, the view that the Supreme Court's ruling about abortion in *Roe v. Wade* applied to pregnant Jehovah's Witnesses who refuse blood transfusions. In retrospect, I wonder how I failed to see the ethical difference between a woman seeking to terminate a pregnancy and a woman who chooses to exercise her right to refuse an unwanted medical treatment.

The minutes of that month's ethics committee meeting contained the ill-fated prophecy: "The final version of the Jehovah's Witness transfusion policy was circulated for review. It will be submitted to the hospital administration following this last survey."

Percolating in the background during the committee's ongoing discussions about the Jehovah's Witness policy was the issue of the necessity and desirability of resorting to the courts. The issue arose in connection with children of Jehovah's Witnesses, as well as with pregnant patients who refuse transfusions. A lawyer-ethicist laid out for the committee the general problems of physicians using the courts to enforce treatment decisions over patients' objections, including the risk of converting physicians into agents of the state and the chilling effect on patients' free use of hospitals due to fear that their wishes may be overridden.

In the case of children, appealing to the courts for required transfusions has the merit of demonstrating due regard for wishes of the refusing parents, wishes that should not simply be ignored by the independent judgment of physicians. The law is clear and specific about the limits of parents' rights to act in a fashion that endangers the life of their children. On the other hand, appealing to the courts to override refusals by pregnant women is a different matter. The law does not impose requirements on pregnant women, nor can it realistically do so. Women should not be forced to violate their own beliefs or to relinquish control of their persons and destiny.

When these issues were laid out before the whole committee, there ensued a lengthy, intense, and animated discussion. A consensus developed in favor of deleting in its entirety the clause specifying a special policy for pregnancy. Instead, pregnant women would be treated in the policy just like other adults with decision-making capacity. Some physicians at the meeting, apprehensive about how their obstetrical colleagues would view the policy, suggested that representatives of the obstetrical service be invited to attend the next committee meeting.

The following month, as planned, two senior members of the obstetrical service presented their views to the committee and stated how they believed their colleagues would respond to the proposed policy. They revealed what they normally do and would continue to do when they judged a pregnant patient to be in need of a blood transfusion. Both said that in cases of an immediate and unmistakable hazard to the fetus, they would seek a court order to transfuse the pregnant woman. They were then asked what they would do if the patient refused to go along with the court' order. Would they forcibly restrain, anesthetize, or otherwise physically coerce a resisting pregnant patient? One physician said "yes" to this question, and the other replied that while he himself would not employ coercive measures, he would respect his colleagues who would.

In light of these sobering revelations, the chairman invited several committee members to make recommendations at the next meeting about what the policy should and should not address. It had become clear that the "final" draft was anything but that, and a rethinking was called for.

At the next monthly meeting, a psychiatrist, an internist, a lawyer-ethicist, and I made brief presentations. Among the issues addressed were these:

Whether court orders should be a mechanism specified in the proposed policy for coercing a pregnant woman into an unwanted transfusion.

Whether a separate section specifying discriminatory treatment for pregnant women was desirable.

The need to include a statement requiring full disclosure by the physician of his or her personal reservations and the likely course of action under circumstances of disagreement with the patient.

The need to provide alternatives to physicians who face this dilemma.

Following a lengthy discussion, the committee affirmed its consensus reached at the earlier meeting to delete from the policy the section specifically devoted to pregnancy. It was better to omit the entire discussion, we believed, than to open that can of worms. The committee also agreed that the policy should not specify the court order option, since it was already available to physicians who chose it. Finally, it was decided that the policy should include a "full disclosure" requirement in a revised, broad introductory section, with guidelines for physicians who could not abide by the patient's wishes. The minutes of the meeting summarized the reasoning behind these decisions:

1. A specific section of the policy would put pregnant women into a special category and could be seen as compromising their rights.
2. Pregnant women have the same rights to control their bodies as do others. The dilemma arises out of the demand that they yield those rights because of pregnancy.
3. It is morally questionable for a woman to refuse a relatively safe minor procedure and thereby jeopardize the fetus.
4. But the religious beliefs of a poorly understood minority demand protection. These rights should be infringed upon only under extreme circumstances.

5. Physicians have an obligation to disclose to a patient their unwillingness to honor the patient's autonomy and to indicate what actions would or would not be taken under specified circumstances.
6. Physicians have a right and an obligation to turn the care of such a patient over to another caregiver if the patient's needs are incompatible with the ethical and professional values of the physician. This is a basic, well-established element of the existing canon of ethics.

The consensus apparently reached at that meeting was only apparent. A member of the committee sent a letter to the chairman a few days later. The letter writer expressed a "cry of anguish," noting that the committee had completely reversed itself over a period of a few months. The letter said:

> . . . we have been decisively influenced by the view of two or three committee members. They are forceful, they are articulate, and I too am persuaded that the outcome they support (acquiescence to the patient's views) would be the right one, at least in most cases.

I recognized myself as one of the two or three "forceful" committee members defending the autonomy of pregnant women. The letter went on to observe that just as the committee's present view is ethically grounded, so too is the contrary view. The writer questioned the adoption of a policy that stipulates, in effect, that it would be unethical to seek a court order authorizing a lifesaving transfusion for a pregnant Jehovah's Witness.

The committee member's letter reflected on the role of an ethics committee, making the thoughtful observation that we cannot simply tell hospital staff "the answer":

> We cannot expect staff members simply to defer to our authority, nor, I think, should we want them to; that would not be ethical on their part. I therefore think that any statement we make on blood transfusions for Jehovah's Witnesses should point out that special issues exist when a transfusion is needed to save the life of a pregnant Jehovah's Witness; it should summarize the competing ethical views; and it should disclose all relevant facts, including the possibility that a court order could be obtained authorizing such a transfusion. . . . If we are not prepared to do this, then I think we should issue no policy on blood transfusions for Jehovah's Witnesses at all.

The committee member's letter was distributed to the entire committee, along with a newly written preamble to the policy and a newly devised section on pregnancy prepared by the drafting committee. The chairman's covering memo said: "At the next meeting we will address the two submissions and decide, given acceptable language, whether we want one, the other, or both, in the best Solomonic tradition."

The new preamble read as follows:

> Early in the course of treatment, a physician must discuss with each patient any religious, ethical, or medical beliefs of the patient that may limit his or her acceptance of a recommended course of treatment. If a physician feels that he or she would be unable, should need for such treatment arise, to accept and uphold the patient's choice, the physician should disclose this reluctance to the patient. The physician should also inform the patient of the alternative courses of action if patient–physician conflict appears likely, including finding another physician within the institution to care for the patient, or even locating a physician outside the institution. As soon as the possibility of actual disagreement becomes clear, the physician should make a good faith effort to locate a doctor who can in good conscience accede to the patient's wishes. If at all possible, this action should not await a crisis situation.

And the proposed new section of the policy devoted to pregnancy, drafted to reflect the discussion that took place at the previous month's meeting, read:

> A pregnant woman has the right to control her body. A dilemma arises out of a demand that she yield this right during pregnancy because some physicians cannot ignore the presence of a "second patient."
>
> Though it is morally questionable for a woman to refuse a relatively safe and minor procedure and thereby jeopardize the fetus, it is also morally reprehensible to infringe upon the person, privacy, and/or firmly held conviction of an informed adult.
>
> Because of the dilemma that arises out of these opposing interests, physicians have an obligation to disclose from the outset their inability to honor a patient's autonomy and must inform the patient of those actions that would or would not be taken under specified circumstances that might conflict with the expressed wishes of the patient.
>
> Every physician has the right and the obligation to try to turn the care of such a patient over to another caregiver if the patient's wishes are incompatible with the physician's professional and ethical values. This course of action is ethically superior to the coercion of an unwilling patient.

At the next monthly meeting of the committee (by this time, 16 months had elapsed since the hospital administrator had requested the committee to draft a policy for Jehovah's Witnesses), three possibilities were considered for the final form of the policy:

1. A policy that made no mention at all of the pregnancy issue and did not address it directly.
2. A policy with a preamble that addressed the issue of conflicting patient–doctor interests.
3. A policy including a specific section on pregnancy.

After much discussion, a consensus formed around the latter two possibilities, with a preference emerging for the third. The committee thus accepted the argument that a policy must address the dilemma posed by the conflicting interests that arise in the case of pregnancy. In the hope of completing work on the Jehovah's Witness policy in a timely manner, the committee agreed to review two alternative model policy statements and choose one. At the next meeting, the committee elected to include both the new preamble and the specific section on pregnancy.

The minutes of the subsequent meeting included a reminder to the committee: "The Jehovah's Witness Policy has reached a 'final' stage and will be submitted for formal approval at the next meeting. Please be prepared to vote on approval." When it came time to vote, the committee unanimously approved the policy and turned its attention to other long-awaited business. Before being implemented, the policy still had to be approved by the medical board of the hospital.

The policy was placed on the agenda of the medical board at first for discussion but not for a vote. The board's response was generally favorable, but predictably, a heated discussion ensued over the procedures for taking care of pregnant Jehovah's Witness patients and, somewhat less predictably, for caring for the children of Witnesses. General concerns were voiced that the hospital could not realistically be expected to provide alternative physicians for those doctors who find a patient's refusal of transfusion incompatible with their own moral standards or feelings. Regarding minors, some board members expressed concern that "mature" children who were not technically emancipated would be prohibited, according to the policy, from refusing transfusion. The majority, however, concurred with the policy's avowedly paternalistic stance regarding minors: They stand in need of protection.

As for pregnant Witnesses, the same set of issues that had been discussed at length during the committee's deliberations surfaced at the medical board's meeting. The chairman of the board, who happened to be one of the obstetricians who had come to speak to the ethics committee, remarked that the policy was inconsistent: It treated infants and children one way and fetuses in the opposite manner. Three committee members who were present at the board meeting insisted that this was not inconsistent but ethically justifiable: Infants and children clearly have moral standing, while the moral standing of fetuses is at best questionable. The committee members defended the rationale that had prevailed in the "final" draft now under deliberation. Further discussion and a vote on the policy were deferred to a future meeting of the board.

When the medical board again took up the Jehovah's Witness policy, two obstetricians came to present their concerns (one was the chairman of the medical board, who had met with the committee a year earlier). Both obstetricians observed that many clinicians could not comfortably treat a healthy, near-term fetus as anything other than a second patient. As before, some members of the ethics committee concurred in this view, while others understood the physicians' feelings but disagreed with their conclusion. One committee member strongly defended the clear assertion in the proposed policy that the woman's rights take precedence over the interests of the fetus. She noted both historical and current political reasons for incorporating this statement into the document. Because of the potential for misapplication to the issue of abortion, the committee had chosen language that avoided identifying the fetus as a second patient or ascribing any independent rights to the fetus.

Now, however, once again deciding to reverse its previously reached consensus, the committee agreed that the unequivocal assertion of maternal rights was too extreme to be practically workable. In the view of some members, it was not even completely ethical. But the committee was unwilling to move to the opposite extreme of embodying in the policy an endorsement of compulsory blood transfusions for pregnant Jehovah's Witnesses.

After two years of work on this policy, the committee had reached an impasse. It became clear that specific guidelines for pregnant Jehovah's Witnesses could not be formulated. Committee members identified two apparent reasons for the impossibility. First, when it comes to weighing the relative interests or rights of the woman and the fetus, the stage of pregnancy inevitably influences the balance. The solution of opting for either extreme—respect for the rights of the woman or the fetus—was unacceptable to everyone. Second, and more important, the problem of ascribing rights and determining their priority has not been resolved by the larger society. It is unreasonable, therefore, to expect a procedurally oriented policy to resolve a dilemma for which no substantive ethical solution has been forthcoming.

Reluctantly, the committee adopted the suggestion that it not even try to dictate a policy but limit its task to describing the competing principles, leaving the decision-making process to the patient and clinician. The committee agreed to that strategy but remained adamant that the policy should state the physician's obligation to disclose to the patient any difference of opinion concerning the woman's refusal of transfusion as soon as such disagreement is evident.

The policy now needed a new paragraph devoted to pregnancy. The group agreed that the section should be rewritten in a way that would not attempt to establish specific guidelines but would identify the competing interests at stake. At the next meeting, a draft of the new paragraph was discussed and approved. The final content and wording of the section devoted to pregnancy read as follows:

Competent adult patients have the right to refuse medical treatment. This right extends to pregnant woman. However, some members of society assert that the fetus has "interests" or "rights" that compete with the rights of the mother to control her own body. In general, the rights of the mother are clearly acknowledged to take precedence in early pregnancy. As gestation advances, it becomes increasingly difficult for some members of society to ignore the "interests of the fetus." Because of the dilemma that arises out of these opposing interests, physicians have an obligation to disclose from the outset if, under specified circumstances, they would be unable to honor a patient's wishes.

Because society and the law have not resolved the conflict between fetal and maternal interests, the policy cannot establish clear guidelines for action where a clinician's interpretation of the interests of the fetus are in conflict with the wishes of the mother. The clinician and patient, with ethical consultation, must seek to resolve such conflicts within the context of the doctor–patient relationship and resort to other means for conflict resolution, including hospital administration, when necessary. Every physician has the right and the obligation to try to turn the care of such a patient over to another caregiver if the patient's wishes are incompatible with the physician's professional and ethical values. This course of action is ethically superior to the coercion of an unwilling patient.

At the next month's meeting of the hospital's medical board, after a brief discussion, the final version of the Jehovah's Witness policy was passed unanimously. The policy was then referred to the hospital's corporate governing body for implementation.

The difficulties that emerged in our ethics committee's effort to develop a policy for Jehovah's Witnesses were, I believe, more idiosyncratic than typical. The idiosyncrasy lay in the fact that the policy had to address the problem of treatment refusal by pregnant women, and that touches issues about which our society is deeply divided. Our ethics committee reflected the larger societal disagreement about these questions and therefore could come to no resolution of the conflict.

One colleague, a physician-ethicist, who is not a member of this ethics committee but who chairs an ethics committee at another hospital, is critical of the outcome. Arguing that this so-called policy is useless to a clinician, he contends that policies ought to help direct action but that this one fails to do so. He says that an ethics committee has the option of issuing a "white paper" that points out various approaches to a problem and outlines competing arguments. But if the task is to devise a policy, the result must be action-guiding.

Should we conclude that because members of the committee held widely disparate views about respect for the autonomy of pregnant women, any attempt to devise a policy must therefore be flawed? What should be the solution where a strong consensus does not exist? Should it be to leave policymaking in such cases to one individual authority who probably holds a single, consistent view about the matter? That has the virtue of efficiency, but it also carries the distinct danger of a dogmatic imposition of one viewpoint about an issue on which reasonable people disagree and where society as a whole remains in conflict.

I agree with my colleague that the Jehovah's Witness policy failed to specify a precise course of action in the case of pregnant patients. The only other option, given the committee's inability to reach a consensus, would have been to omit reference to pregnancy altogether. But that option, too, had been tried and rejected. This cautionary tale also illustrates a dilemma for ethicists who serve on committees. Adhering to the arguments I have made in this book, my position on the issue of pregnant Jehovah's Witnesses, from beginning to end, supported the right of the competent adult patient to refuse treatment. However, since working on a policy in a committee is a cooperative venture, in the end I capitulated in order to help forge a consensus with colleagues.

The same physician-ethicist who criticized this section of the policy as useless chastised me for contributing to the wishy-washy outcome. Where, he asked, was my "minority response" to the policy? I think my lack of response stemmed not from moral weakness but from a feeling of powerlessness. Ethicists are recognized as consultants and teachers in the biomedical world, and for their expertise in analyzing and clarifying thorny issues. But we lack the traditional authority of the physician in the hospital, and ethicists and physicians alike lack power when it comes to confrontations with hospital administrators.

A postscript to the saga of the ethics committee's long and arduous work on the Jehovah's Witness policy does not reveal a happy ending. Lawyers for the hospital's governing board refused to endorse the policy, primarily because of its statement about pregnant women. The official legal position of the governing board requires that the hospital administration go to court for a judicial ruling each time a pregnant patient refuses a treatment by a physician when there is an assessment of possible harm or death to the fetus. The attorneys were unswerving in their insistence on adhering to that practice. In a way, the final draft of the policy gave them no choice. If there is no consensus, lawyers will insist on seeking a court's decision.

Ironically, this same governing board rejected the necessity for obtaining a court order in cases where Jehovah's Witnesses refuse blood transfusions for their minor children. Although that had long been the practice of the hospital and is the prevailing practice elsewhere, the governing board now stated that it would be permissible just to go ahead with the transfusion, since courts invariably follow the physician's recommendation and order the blood transfusion. The reason behind the governing board's reversal of policy? It costs money each time the hospital seeks a court order, so avoiding that step will save money.

In response (perhaps also ironically), the ethics committee objected to the governing board's decision *not* to go to court in cases involving children. The basis for the committee's view was that respect for a minority religious sect requires adherence to due process of law. The ethics committee unanimously supported the substantive ethical position of pediatricians, who believe their obligation to their young patients demands that they advocate for their patients' best interests. Since children do not possess full autonomy that could enable them to decide for themselves whether to accept or refuse blood, they need advocates to protect them, even against decisions made by their own parents. But if doctors simply ignored the child's parents and administered the transfusion, that would ride roughshod over the rights ordinarily accorded parents to consent to medical treatment for their children. To seek and obtain a court order makes the state, not the physicians, the agent forcing parents to accept an unwanted treatment for their child.

This difference may appear to be only symbolic, since the outcome is the same in either case. But from the point of view of the physician–patient relationship, the difference is significant. Physicians should not be put into an adversarial relationship with the parents of their pediatric patients. It surely works to the advantage of children when their parents and physician are allies in promoting their health and well-being. Jehovah's Witnesses typically regard an order issued by a court as an action against their will and over which they have no control. The state is a powerful entity. In contrast, the ethics committee reasoned, physicians should not loom as powerful entities authorized to disregard parental wishes.

The principle known as "respect for persons" usually dictates that a patient's wishes about medical treatment be honored. In this situation, refusing to honor the wishes of Jehovah's Witness parents and acting in the best interests of their young patients is the ethically right action. But physicians can nevertheless show respect for the parents by adhering to the requirements of due process, making the state rather than the doctor the agent of coercion.

The question remains whether resorting to the courts might be avoided by putting in place a review process within the hospital. Since it is a foregone conclusion that judges will order transfusions for children of Jehovah's Witnesses, calling a judge is not a legal necessity. Instead, the ethics committee might be called on to review such cases. This could serve the dual purpose of adhering to the requirements of due process and preventing pediatricians from being thrust into an adversary position with the parents of their patients. An added advantage is that parents could be invited to come before the ethics committee in person. Not only would they have an opportunity to state their views, but they could also hear members of the committee voice their sincere convictions in advocating for the life and health of pediatric patients.

Ethics Committees and Hospital Administration

Happily, the sorts of conflicts that typically come within the purview of an ethics committee devising policy do not usually reflect deep divisions within society. Yet other tensions often arise. The hospital's interest in minimizing liability or costs and the ethical obligation to respect the rights of patients may always be a source of conflict. The wishes of some paternalistic physicians to continue practicing the way they did in an earlier era will no doubt provide an ongoing source of conflict needing resolution through policy development (recall the group of rude anesthesiologists who called respecting the rights of Jehovah's Witnesses "murder"). Although these issues may pose difficulties for an ethics committee, they are not the sort that reflect deep divisions in the moral views of citizens in the larger society. Nevertheless, they remain typical of the conflicts committees do and will face in their work on hospital policies.

Are ethics committees independent of the hospital administration? Should they be? One criticism of ethics committees can be expressed in the following argument. Either an ethics committee is appointed by and reports to hospital administration, or it does not. If the committee is part of the hospital administration, it cannot be independent and is bound in its recommendations and actions to reflect the administration's point of view. If the committee is not part of the hospital administration, it will lack the power and authority necessary for its recommendations and decisions to be binding. In either case, hospital ethics committees will be ineffective in ensuring that ethically sound, patient-centered decisions or policies are implemented.

This argument raises a question about whether ethics committees do, in fact, accomplish their mission of promoting the interests and protecting the rights of patients. Some critics doubt that these noble purposes are typically served. One of these skeptics, George Annas, argues that committees are probably least suited for the task of consulting on individual cases. Such consultations almost always end up being dominated by "the lowest common denominator the law allows" rather than dealing with "higher" ethical concerns. Moreover, Annas sees committees as probable tools of the hospital administration: "Insofar as their primary mission is to protect the institution by providing an alternative forum to litigation or unwanted publicity, the term 'ethics' is inappropriate, and the committee should be called a 'risk management' or a 'liability control' committee."[3]

I am forced to agree with Annas's assessment in situations where the hospital administration has an active voice in the operation of the committee. It is undeniable that risk managers often stand in the way of recommendations by ethics committees and subsequent performance by medical staff of the committee's recommended action. Such arrangements raise serious questions about how well the purpose and function of ethics committees are served. All too often, the committee can act as a vehicle or even a rubber stamp for the hospital's administrative interests. Yet at a conference I attended recently with a group of leading philosophers, lawyers, and physicians working in the field of bioethics, several participants mentioned that the risk manager and in-house counsel were not allowed to serve as members of their hospital ethics committees.

These committees are now becoming institutionalized, with the result that a wide array of the institution's interests are being pushed onto their agenda. Is that always an evil to be avoided? Not if having the institution's interests on the agenda provides benefits to patients. As institutions, hospitals have multiple agendas, some of which are designed to enhance their own image and to promote their financial well-being as much as to serve the sick.

It is not the responsibility of an ethics committee to minimize hospital risk. There are plenty of other players in that game. And it is surely not the ethics committee's business to become involved in allocating economic resources in the most efficient manner. Not only would that activity be beyond the expertise of most ethics committees, but it is also beyond their purview since it represents financial interests and not ethical concerns. However, there may be a role for ethics committees in helping to allocate the hospital's resources in an *equitable* manner. Justice in the allocation of resources is an ethical concern and one that is patient centered. How good ethics committees are at devising equitable allocation schemes remains to be seen.

Conflicting views about the proper object of their concerns have provoked ongoing debate over the question of whether ethics committees should take institutional interests into account when making recommendations or devising policies. How could a committee not take institutional interests into account when devising a policy for the institution? If ethics committees ignore institutional interests, they are impractical and unrealistic. Ethical concerns are only one among many that complex institutions like hospitals must deal with.

An administrator might argue as follows: "The administrator must consider such issues as financing and community attitudes because doing so is a part of the administrator's responsibility, and because the administrator's authority covers not only the case or policy choice before the ethics committee, but the full range of cases and policies in the institution."[4] If an ethics committee is to serve the institution, it must take administrative concerns into account.

On the opposite side is the purist view, contending that the job of ethics committees is to determine "what is right" rather than "what is feasible."[5] It is better for the ethics committee to put forth purely ethical judgments since the committee will most likely not have the last word in the matter. It is up to the hospital administration to weigh the ethical determination against other institutional concerns in making the ultimate decision regarding the case or policy.

I strongly endorse the purist view. When our ethics committee presented its version of the Jehovah's Witness policy to the medical board of the hospital, one physician asked whether the committee had sought advice and counsel from lawyers from the corporate governing board. We replied that we had not, but one of the committee members is a well-known academic attorney, thoroughly familiar with relevant health law. We told the medical board that we knew the corporate board's lawyers would have to review this policy before it could gain final approval, so there was no need to include their input at an earlier stage of the process when the focus should remain on the ethical issues.

Although the policy was flawed in its failure to give clear guidance to caregivers of pregnant Jehovah's Witnesses who refuse transfusions, if we had consulted lawyers from the governing board at an early stage, they would only have insisted at the beginning of the process what they mandated at the end: The court must be petitioned in all cases. From a practical standpoint, there is admittedly no difference. But the process of deliberation, of forming and reforming a consensus, was instructive. On moral matters about which reasonable people disagree, the question of whether it is possible to devise a policy that gives clear guidance remains unanswered.

In a similar vein, I think committee members who sided with hospital administration in the case of the Holocaust survivor were inappropriately taking the institution's interests to be part of the ethical analysis. Here it is important to distinguish two different positions those supporters might adopt. The first position is that increased costs to the hospital in continuing to provide a high level of care for the patient is one of the *ethically* relevant considerations. This position holds that physicians' obligations to their patients (or families) must be balanced against obligations to others: to society, to the institution, or to the insurance pool.

The second position supporters of the administration could take is to deny that financial costs to the hospital are part of the ethical analysis of the case but nonetheless to affirm that it is proper for the ethics committee to take institutional interests into account. This latter position rejects the purist view of the ethics committee's role but still distinguishes genuine ethical issues from financial considerations.

There is a third position, which casts the ethics committee in a different role, mentioned earlier: that of helping to allocate the hospital's resources in an *equitable* manner. If other patients are adversely affected by the resources being consumed by the Holocaust survivor who will not recover her mental function, then there is an ethical argument for reallocating those resources. When a permanently unconscious patient is occupying a bed in a critical care unit that is needed for another patient who could derive greater benefit, then the ethics of triage can support lowering the level of care for the unconscious patient. If the resource in question is money rather than a hospital bed in a critical care unit, then the hospital must be able to show that the money saved on this patient will actually be used to care for other patients who would not otherwise receive that care. Justice in the allocation of resources is certainly a patient-centered ethical concern, but one that must simultaneously take into account the interests of more than one patient.

As they have become more numerous and more entrenched, ethics committees have received their share of criticism. Some doctors have objected to the intrusion of yet another outside force into the physician–patient relationship, a legitimate ethical concern that needs to be guarded against. Voicing a related worry, other doctors have lamented further erosions of their own autonomy and professional authority. A difference, though subtle, marks these two concerns. The first objection focuses on the physician–patient *relationship* and its ethical aspects, which include preserving confidentiality and maintaining trust. The second objection focuses on the *professional role* of the physician, arising out of concerns characteristic of all professionals: the desire to maintain independence and control.

Even staunch defenders of patients' rights have been skeptical of ethics committees. George Annas, who has long been a champion of patients' rights, acknowledges his growing skepticism: "'Ethics committees' have grown from an anomalous entity to provide ethical comfort to a few, to an almost standard entity to provide ethical cover for many."[6] Even worse, Annas contends that the primary preoccupation of committees "has been to respond to legal changes and challenges, rather than to do anything a philosopher might label 'ethics.'"[7] These criticisms are probably accurate statements of a role many committees have adopted, intentionally or unwittingly. If so, the critical charges are warranted. But the question remains, must ethics committees inevitably fit this description, or might they aspire to loftier goals?

In coming to the defense of ethics committees, I endorse what I take to be their proper role, not the way many committees probably operate in actual practice. I do not contend that all ethics committees have adopted that role, or even that those committees choosing to act as advocates of patients succeed in achieving that aim all the time. Again, the distinction between ethics as an ideal and ethics in practice needs to be emphasized. At best, a well-functioning, well-constituted hospital ethics committee has only a modest role to play. But if it takes seriously its role of advocating for the rights and interests of patients, it may help to stave off the enemies.

Notes

1. John D. Lantos, Steven H. Miles, and Christine Cassel, "The Linares Affair," *Law, Medicine & Health Care*, vol. 17 (Winter 1989), p. 309.

2. The account that follows is adapted from my article, "The Inner Workings of an Ethics Committee: Latest Battle over Jehovah's Witnesses," *Hastings Center Report*, vol. 18 (1988), pp. 15–20.

3. George J. Annas, "Ethics Committees: From Ethical Comfort to Ethical Cover," *Hastings Center Report*, vol. 21 (1991), p. 19.

4. Daniel Wikler, "Institutional Agendas and Ethics Committees," *Hastings Center Report*, vol. 19 (1989), p. 22.

5. Ibid.

6. Annas, "Ethics Committees," p. 18.

7. Ibid., p. 19.

11

The Decline of Autonomy

Most physicians care about their patients and want to do the right thing for them. Yet it is an inevitable fact of everyday life, as well as the practice of medicine, that the right thing to do is not always perfectly clear. Ambivalence or uncertainty about the right course of action in an increasingly complex world of medicine is one of the factors that contributed to the emergence more than two decades ago of the field of bioethics. In medicine today, physicians continue to confront old and familiar dilemmas, but sometimes with a new twist. I am surprised at many of the beliefs and attitudes of the emerging generation of physicians.

An example is the apparent shift in attitude regarding the physician's duty to preserve confidentiality and the accompanying belief that physicians are as much the agents of society as they are of the patients who come to them for care. The most common situations in which tension exists between duty to the patient and obligations to society are those of psychiatric patients judged to be dangerous to others (Chapter 6). The AIDS epidemic has produced renewed debate within and outside the medical profession about when obligations to the patient are overridden by public health considerations, a debate that was vigorous in the days before most infectious diseases had been conquered.

The shift in attitude regarding physicians' obligation to preserve confidentiality was evident in a recent conference in clinical ethics for medical students. I co-teach this conference with two physicians, one a professor of medicine in his middle to late fifties and the other a younger doctor in his mid-thirties. The case presented at the conference, which produced heated discussion and sharp disagreements, centered on a 22-year-old man who was found on the pavement and brought to the hospital emergency room by the police. Diagnosing his condition as an apparent drug overdose, physicians in the emergency room gave the man medication to counteract the effect of the drugs. The patient woke up combative and violent. He informed the emergency room staff that he had taken a combination of large amounts of drugs and alcohol. Now he wanted to leave the hospital and go directly to work.

The man disclosed the fact that he drove a truck for the city's Department of Sanitation. In fact, if he left the hospital to go to his job, he would be driving a truck all night. The medication he had been given to counteract the drug overdose clears the body in two to three hours. He was in no further danger from the drugs and alcohol he had ingested, and his mental status was judged to be satisfactory, so there was no medical reason to keep him in the hospital. Despite this, the emergency room physicians were reluctant to let him go. They required him to sign a form stating that he was leaving the hospital against medical advice, (AMA), and the patient left.

The question posed in the emergency room was whether the doctors should call the patient's employer and disclose that the man had been consuming large amounts of drugs and alcohol and was brought to the emergency room unconscious. One physician said, "I won't call," but told a medical student she could make the call if she wished. The student agonized over the decision and waited several days before deciding that it was her obligation to call the Department of Sanitation. Without disclosing the purpose of the call, the student succeeded in reaching the patient's supervisor and learned that the man had already been fired. She did not inquire about the reason for his dismissal.

At the conference, a number of vocal students supported the action of this student. Some were prepared to go even further, saying that there must be some way to keep patients like this in the hospital in order to protect the public. Analogies were made between this patient and psychiatric patients who engage in violent behavior and are deemed dangerous to others. Such patients are detained until they are considered no longer a threat to others. It was suggested that even if this patient suffered loss of employment as a result of breaching confidentiality, it was justifiable because he shouldn't have a job where he might endanger the public by driving under the influence of drugs and alcohol.

On the other side, it was pointed out that there was no clear evidence that this patient drank or took drugs on the job. Maybe he just went on weekend binges. Perhaps he was a regular user but always sobered up before going to work. Those who argued that the man was a danger to others were merely surmising that he drank or took drugs routinely before going to work or while on the job. Since he was sober at the point when he insisted on leaving the hospital, he was not at that moment a danger to others.

Students in the first group hardened their position. "Look, this guy's a druggie. He's a menace to society. We're physicians, and we have an obligation to protect the health of the public. If we know something about a patient that threatens the health and safety of the public, we are obliged to do something about it."

"Don't we also have an obligation to protect the confidentiality of our patients?" someone on the opposing side asked.

"Yes, in general, but not always. This is a good example of the exception to doctor–patient confidentiality. Besides, this guy was brought to the emergency room of a city hospital by the police. There's no real doctor–patient relationship here. He didn't seek out a relationship with us. It would be different if he came to a physician's private office. Then we'd have a doctor–patient relationship and a duty to preserve confidentiality."

A spokesperson for the opposing group replied: "But that would be unjust! You're saying that patients who go to a physician's private office are entitled to have their confidentiality protected, but patients who can only go to the city hospital are not."

"That's not really the point. The point is the danger to innocent persons. Look, the police are authorized to pick up drunken drivers. Don't we have a similar obligation to prevent drunken driving if we know about it?"

At this point, there were enough issues on the table to go on for days. The senior physician at the conference took a straw poll, asking two questions: How many thought breaching confidentiality was justified in this case? How many thought it was not? Most of the students in the room raised their hands in response to the first question. Three students were opposed to calling the patient's employer, and the younger faculty teacher agreed with them. The senior physician and I did not vote on either side, which irritated one of the students. I replied that I refrained from voting only because I didn't want to influence any of the students, but they would surely hear my view when I summed up at the end of the conference. The senior physician acknowledged that he believed that the obligation to preserve confidentiality was not absolute, but in this case it superseded some vague duty to the public.

Animated discussion continued, and at the end, I concluded with my usual summary of the issues. Along with an ethical analysis, I stated more fully the argument in favor of maintaining confidentiality. Yes, there are circumstances in which breaching confidentiality is warranted. But those are situations where the probability of harm to others is reasonably high and where the degree of harm is significant. No one knew the probability that this patient was actually in the habit of driving the sanitation truck while drunk or after taking drugs. It is hard to assess the likelihood or severity of harm even if this man did drive while under the influence, since he drives a slow-moving vehicle in nearly deserted city streets in the middle of the night. There is simply not enough evidence regarding the likelihood and severity of harm to justify breaching the well-entrenched duty of physicians to keep confidentiality.

Why does that duty exist, and why is it important? In virtue of their profession, physicians are granted awesome authority to inquire into the most intimate details of their patients' lives. No question is too prying to be legitimate if it bears on the diagnosis or treatment of a patient. The normal social boundaries of privacy are virtually absent in the physician–patient relationship. Even when no relationship exists, people think nothing of talking to a physician about personal matters. The younger physician at the conference said he was amazed at the things casual acquaintances and even perfect strangers told him when they learned he was a doctor. To test this hypothesis, he once decided to do an experiment. While at a party engaging in social conversation, he first told people he was a doctor and then casually asked about their bowel habits. With no reluctance and no visible embarrassment, virtually everyone who was queried answered the question.

If patients are to feel free to disclose such intimate details (short of an intention to harm another human being), they must feel confident that what they tell their doctor will go no further. Trust in the particular doctor a patient is seeing, and trust in the medical profession generally, are essential for the provision of good medical care. If people constantly had to worry that what they told their doctor might be disclosed to others, including persons like employers who could harm their interests, it could very well inhibit the imparting of information critical for diagnosis and treatment.

In one respect, there was nothing new about this discussion. Over the years I have worked in the field of bioethics, thoughtful people have come down on both sides of hard cases that pose dilemmas of confidentiality. Yet in another respect, there was something new and very troubling about the view of the majority of the students at this conference. Several of them insisted on making analogies between their role as doctors and the role of a police officer: "The police can detain someone who's drunk"; "The police can stop motorists if they appear drunk"; "The police can involuntarily hold someone who appears to be a danger to others."

I find this analogy between physicians and the police inappropriate and distressing. Police are official agents of the state, while physicians are not. Physicians are supposed to be agents of their patients, respecting their rights and advocating for their interests. To analogize the duty of physicians to that of the police in protecting public safety is, at the very least, to invest the role of physician with divided loyalties. At worst, it makes the physician a perpetual double agent, always having to vacillate between obligations to patients and obligations to a vague, undefined public.

One caveat is in order. There are certain contexts in which physicians are employees of institutions that do command their loyalty in ways that have built-in problems of double agency. Psychiatrists who work in prisons or mental institutions, physicians who serve as directors of health care facilities, doctors in the military and those who do examinations for insurance companies, and perhaps also physicians who work for a college or university health service all owe loyalty to the institution and its mission, in addition to having obligations to the patients they treat. In those contexts, the obligation to preserve the confidentiality of patients may well be lessened, but the obligation to be truthful is not. Physicians who work in situations where their loyalty is truly divided have the obligation to inform patients of the limits of doctor–patient confidentiality. In the absence of a warning that the usual dictates of confidentiality do not apply, patients have reason to believe that what they say to a physician will be held in confidence.

Do patients brought to a hospital emergency room have reason to believe that the usual dictates of doctor–patient confidentiality apply? I contend that they do. Again, there are exceptions. If a patient is brought to the emergency room with a gunshot or stab wound, physicians have a legal duty to inform the police. If a patient brought to the emergency room tries to leave AMA, stating that he was injured while committing a crime and must now go to finish the job, physicians may detain him and notify the proper authorities. Wouldn't a patient have to be crazy to disclose that he was in the midst of committing a crime? Perhaps. The absurdity of that scenario suggests that only someone who is mentally impaired would make such a disclosure.

But now recall that the patient presented at the ethics conference openly stated what his job was and where he worked. He would never have given that seemingly innocent information to doctors if he suspected that they would disclose to his employer that he had been hospitalized for an overdose of drugs and alcohol. He must have believed, implicitly and quite appropriately, that even these doctors he was meeting for the first time in the emergency room of the city hospital had a professional obligation to preserve the confidential information he provided.

When the conference was over, I asked both the senior physician and my younger faculty colleague whether they thought medical students in the past would have reacted the same way as this group had, and whether something is changing in regard to young physicians' loyalty to and advocacy for patients. Both physicians expressed surprise at the way the straw poll had gone, and also at the strong sense in which these medical students felt themselves to have police-like roles. The younger physician said he couldn't put his finger on what was disturbing him throughout the discussion, but once I mentioned it, he realized how alien it was to him to think of members of his profession as agents of the state.

Several students lingered after the conference, and one noted that he grew up in the anti-drug era. Everything he heard in school and at home was so infused with loathing and mistrust for drug users that patients who use drugs are viewed by him and his fellow medical students as a special class of undesirables. To which the older physician replied: "This was outside my experience as a medical student and young physician, since I grew up in the predrug era." Then, pointing to our other colleague, 10 years older than the average current medical student, the senior physician observed that the younger doctor had grown up in the "pro-drug era." These sociological observations are interesting, but I believe they have little to do with the ethical obligations owed by physicians to their patients.

Even when doctors feel unquestioned loyalty to their patients, forces external to the doctor–patient relationship increasingly prevent them from being unswerving advocates of their patients' interests. It is evident that hospital administrators, risk managers, and the lawyers who work as hospital counsel have multiple interests, some of which ignore or even conflict with patients' rights and interests. And although the law can guarantee the moral rights of patients, physicians by and large are uninformed about legal details even when legislation or regulation bears directly on clinical practice. A case in point (discussed in Chapter 2) is the law in New York State regarding DNR orders. Despite the fact that the law has been on the books for several years, many physicians remain woefully ignorant about its provisions.

But the DNR law is somewhat peculiar in that it addresses only one medical treatment: cardiopulmonary resuscitation. What about more general laws that seek to empower patients, such as living will legislation or health care proxies? The purpose of these advance directives is to ensure that patients' values and wishes about treatment can be carried over to a time when they have lost the capacity to speak for themselves. A new federal law—the Patient Self-Determination Act—went into effect on December 1, 1991. This statute requires all health care institutions that receive federal funds to provide information to all patients concerning their legal right to make decisions concerning medical care, including refusal of treatment. The laws in each state vary in their details, so the information provided to patients in one state will be different from that in another. But the federal law requires the health care institution to document patients' advance directives in their medical record, and to provide education about these matters both to its own staff and to the surrounding community.

Having information about patients' previously stated wishes should make things easier for doctors and hospital administrators, as well as for patients' families. Risk managers would no longer have to worry (or worry as much) about a family's request to withdraw treatment. If a family member has been appointed as the patient's health care agent, or if the patient's wishes are written down in a living will, the paranoia that leads risk managers to take overly cautious stances (even when everyone else agrees on withdrawing treatment) may abate. Of course, matters could go the other way for a hospital administration seeking to withdraw treatment. If Helga Wanglie had legally appointed her husband as her proxy, or had she specified in a living will that she wanted to be kept alive artificially, her rights would have been more secure.

Nevertheless, many people are skeptical about new laws that threaten to change the way things are normally done. Not long ago, I was on a panel speaking to a college alumni group about advance directives. Also on the panel were a doctor, a hospital administrator, and a lawyer. When it was the doctor's turn to speak, he made a commonsense plea to the audience: Why do we need all these laws? Things were going very well without laws to instruct doctors, nurses, hospital administrators, and patients' families in what they may or may not do. In the "old days" (this physician was only in his early forties), doctors talked to patients, or to the patient's relatives, and they just decided what to do. There was no need to learn the many details of how to implement a law, nor were doctors and patients burdened with having to fill out these numerous forms.

The audience could only take the speaker's word for his assurance that along with other physicians, he succeeded in communicating sensitively and thoroughly with his patients. But many listeners had had an altogether different experience with the medical profession. Having a doctor assure them that patients don't need laws to protect their rights was not a sufficient response to many who had personally encountered uncommunicative or paternalistic physicians.

The lawyer on the panel spoke knowledgeably about the state law—a health care proxy law—but she was stumped on the problem of what to do if a patient had expressed to the nurses a wish about treatment, then fell unconscious, and the appointed health care agent requested something different from what the patient had wanted. Whose views should take precedence? That particular conflict was not directly addressed in the state law, and the lawyer admitted not knowing the answer.

The lawyer was not the only one who was puzzled about this situation. A nurse brought a case to one of the hospital ethics committees on which I sit, distressed about an episode that left her frustrated and angry. The nurse was caring for a patient who had AIDS. When the patient entered the hospital, he was conscious and mentally alert. Fully aware of his prognosis, the patient expressed a wish not to be resuscitated and early in his stay signed a DNR order. He also executed a valid health care proxy, naming his mother as his health care agent. When her son lapsed into unconsciousness, she demanded that everything be done to keep him alive. In fact, the patient suffered a cardiac arrest and his mother, who was then his legally authorized agent, insisted that he be resuscitated. The nurse knew very well what the patient would have wanted and produced the signed DNR order. Yet she was not permitted to carry out the patient's request.

Of course, the hospital's risk manager was called. The risk manager said, "You have to listen to the agent. The mother is the legally authorized person to speak for the patient. That's why the state legislature passed this law, so that there's no doubt about who may decide for an unconscious patient." When the nurse brought the case to the ethics committee, those present were unanimous in their belief that carrying out the patient's wishes was ethically the right thing to do. But most were unsure of how this conflict could be legally resolved.

One knowledgeable committee member insisted that the correct reading of the state's pertinent laws gave the health care agent (the patient's mother) ultimate authority to decide. Anyone who wished to challenge the decisions of a properly appointed agent would have to go to court. The committee member's knowledge derived from his position on the state's Task Force on Life and the Law, the body that issued the report and recommendations on which the health care proxy law was based.

Others at the ethics committee meeting disagreed. Another knowledgeable committee member expressed certainty that going to court is required only when someone challenges the health care agent's decision in cases in which the patient has never expressed any wishes. In the case at hand, however, the patient's wishes were stated in evidence that was surely clear and convincing, in the form of a valid, signed DNR order. As the ethics committee was later to learn, the hospital's risk manager came to believe that her earlier interpretation of the state law was mistaken. She subsequently reversed her earlier statement that the patient's mother had the ultimate authority to decide unless the case was taken to court. Adding to the confusion and uncertainty, an attorney with the state's Task Force on Life and the Law affirmed the risk manager's original position, the same position taken by the committee member who served on the Task Force: The decision of a health care agent is the final word in such cases.

These sorts of conflicts underscore doubts about whether laws help or hinder in securing patients' rights. Although the Patient Self-Determination Act was only recently enacted, numerous concerns have already been voiced.[1] There is considerable evidence that patients' advance directives are often ignored even when they are known. One study suggested that this happens about one-fourth of the time.[2] Another position taken by many doctors is that patients do not really want to discuss these things because they don't want to be reminded of their mortality or of a future time when they might not be able to speak for themselves. As a result of this belief, many doctors fail to initiate discussions with their patients about advance directives. Other doctors simply don't want to take time to talk to patients about matters other than their immediate treatment.

These problems do not arise from any flaws in the laws themselves, but they do show how hard it is to change the behavior of professionals who are used to acting in a certain way. Better education of physicians and patients alike might improve matters somewhat, and it should be the responsibility of medical schools and postgraduate training programs to do more than simply acquaint doctors with the fact that a new law exists. Young physicians (and older doctors as well) can benefit from supervised clinical sessions carrying out discussions with patients about their future care. Just as clinical diagnosis and treatment is best taught by experienced and sensitive clinicians, so too are role models needed for training physicians to approach patients about topics they have formerly considered taboo.

There is another, deeper source of concern about advance directives. It goes back to a basic dilemma highlighted in bioethics: the potential conflict between patients' rights and their best interests. Laws that authorize recognition of living wills and appointment of health care proxies are designed to empower patients. While still competent, people can express wishes about what they want when they are no longer able to speak for themselves. In recognizing this extension of patients' autonomy, such laws transform a moral right of self-determination into a legal right.

But what if the choices stated in an advance directive fail to accord with what everyone would agree is in the best interest of the now incompetent patient? Is it ethically right simply to ignore these current interests? Situations in which this conflict might arise may be few in number, but they are nonetheless poignant. The problem does not come up with patients who are comatose or in a PVS because, arguably, they no longer have any interests that can be served by continued medical treatment. But there are patients who are so demented that they cannot voice their views, yet who might still be able to derive pleasure from continued life.

Experts in bioethics disagree about what is the right thing to do in such situations. Some argue that it is essential that advance directives be honored. Without having the confidence that their wishes will be respected, why should people take the time and trouble to prepare living wills or appoint a health care agent? Furthermore, this argument contends, it is a violation of people's right to autonomy to have a system that leads them to believe one thing but can reverse that presumption at a later time.

Yet the opposing argument is also compelling. It holds that patients in this situation are really different people than their former selves who made the advance directives. Their circumstances have now changed so much that they may no longer hold the values they expressed earlier. People do change their minds about what they find tolerable or acceptable. A younger, active person's strongly held value that she would not want to receive medical treatment were she to lose her mental faculties might change with the perception that simpler pleasures could still make life worth living.

It is not easy to resolve this conflict between respect for autonomy and acting in a patient's best interests. Reasonable people disagree about what is the ethically desirable solution. Unlike the defense of strong paternalism that would override the wishes of competent patients who are still able to speak and decide for themselves, the view that vulnerable patients need protection supports a weak paternalism. My own view is that unless a patient has specifically addressed this situation in a living will, stating unequivocally that dementia of any degree constitutes an unacceptable quality of life, then a good case can be made for continuing treatment. Although I remain opposed to coercing competent patients into accepting treatments that their physicians or families believe are in their best interest, I think it is justifiable in certain cases to treat demented patients differently from their previously stated wishes. Advance directives should not be overridden lightly. The kind of circumstances that would justify ignoring a patient's previous wishes are those in which there is evidence that despite having lost some mental and physical functions, the person is still enjoying some of life's pleasures—tasting food, having a back rub, sitting in the sunshine, recognizing relatives who come to visit.

The Patient Self-Determination Act is an example of a law designed to empower patients, granting people an extension of their right to autonomy. However, even well-intended laws like New York's DNR law have been shown to have unintended consequences. Simply enacting a law cannot guarantee that what it is designed to accomplish will occur in all cases. Unfortunately for patients and doctors alike, the trend in this country has begun to shift away from respect for autonomy. This trend is evident in developments that go well beyond the power vested in hospital risk managers to thwart decisions by physicians and patients. The latest development goes under the name of "managed care."

"Managed care" is a euphemism for transforming medical decision making into business decisions. Health plans that employ managed care features typically do not permit patients who enroll in them to choose any physician they wish. They are limited to physicians employed in a health maintenance organization (HMO) or on a list of "preferred providers." Furthermore, under most managed care plans, physicians who are employees of the HMO or are one of the preferred providers are limited in what they may offer patients. There may be limitations on what tests may be performed, what experimental treatments may be offered, how long a hospital stay is allowed, and whether ancillary treatments such as physical therapy will be available under the plan. Managed care plans severely limit the opportunity of physicians to advocate for their patients. Even if that advocacy is based on expert medical judgment, individualized to the particular patient, the doctor cannot promise what the administrative plan refuses to offer.

Managed care plans are being highly touted as a means of lowering expenses for health care. Yet it is not at all clear that they can succeed in achieving that goal. These plans leave in place the multiplicity of third-party payers, both public and private, that make the health care delivery system so

costly to administer in this country. Designed to lower the costs of diagnostic
tests, treatment plans devised by physicians, and stays in the hospital, managed care plans leave untouched (and may actually increase) the vast administrative bureaucracy required to oversee every minuscule maneuver in the
practice of medicine.

Thoughtful citizens concerned about the costs of health care need to be
wary of politicians beating the drum for managed care. Transforming medicine into even more of a business operation than it already is cannot be the
solution to the problem of escalating costs. More important, managed care
limits further the choices that patients and physicians alike can exercise. It is
undeniable that managed care plans erode the opportunity for doctors to act
in their patients' best medical interests.

From an ethical point of view, it is important to distinguish the different
aims and purposes of the numerous overseers in today's medical practice.
The aim of managed care plans, quite simply, is to keep health care costs as
low as possible consistent with meeting the existing standards of medical
care. To achieve that goal, health care institutions, such as HMOs, and
companies that provide insurance for health care, set limits that doctors must
adhere to and patients must accept. Neither the plans themselves nor the
people who administer them are patient centered. The objective is not to
achieve what is best for the patient but, rather, what is best for the company
or health care organization.

In managed care schemes, bureaucrats make *prospective* decisions about
what diagnostic and therapeutic maneuvers doctors may perform. The insurance company or health care organization refuses to reimburse physicians for
procedures that go beyond what the preordained plan allows. In contrast,
utilization review (discussed in Chapter 7) is *retrospective*. In this system of
chart and record review, overseers look back at decisions or recommendations
made by physicians and determine which ones are proper or allowable.
While it is true that utilization review can weed out incompetent or irresponsible physicians who recommend treatments or hospitalization stays that are
unnecessary or excessive and may risk harming patients, the system can also
restrict physicians and hospitals from providing what may be in an individual
patient's best interest. In order to avoid having to justify an unallowable
treatment recommendation after the fact, some physicians feel compelled to
determine in advance just what the utilization review standards are. So,
although the system operates by an examination of past treatment decisions
and completed actions, the standards developed by companies that perform
utilization review can constrain the actual decisions physicians make on
behalf of their patients. Here again, the goal is to minimize costs by eliminating unnecessary procedures and hospital stays. The goal of protecting patients from bad medical decisions is secondary.

Hospitals employ still another mechanism for reviewing the decisions and actions of physicians and other health care personnel. That mechanism is known as "quality assurance," which is carried out in every hospital department. Quality assurance is essentially a self-policing method designed to identify physicians who are practicing substandard medicine. The practice of bad medicine can be a result of impaired physicians, careless procedures, negligence, or incompetence. The process of monitoring takes place in regular meetings held in departments of medicine, surgery, obstetrics and gynecology, the emergency room, and elsewhere in the hospital. The purpose of the meetings is to review procedures after examining medical charts and records, and thus to identify any incidents or practices that fall below accepted standards of care.

It seems obvious that if the mechanism of quality assurance works as it is supposed to, it can have the desirable result of achieving an adequate if not higher level of care and treatment of patients. However, what drives the quality assurance system in hospitals, by and large, is not the aspiration of administrators or health professionals to deliver the best possible care to patients. Instead, it is the need to ensure that there are no deficiencies that would pose a risk of the hospital's losing its accreditation. The chief organization in the United States that accredits hospitals is known as the Joint Commission on Accreditation of Healthcare Organizations (JCAHO). The organization conducts its work through published standards and by regularly scheduled visits to hospitals and inspection of hospital records.

Skeptics have often criticized systems of self-monitoring, and with good reason. Historically, the medical profession has had a dismal record of weeding out physicians who are impaired or incompetent. Professions not only seek to protect their members from scrutiny and sanctions by outsiders, but also have a stake in maintaining the reputation of the profession as a whole. Only recently has it emerged that the system of quality assurance in hospitals embodies widespread practices of dishonesty and deception. It is dismaying (but perhaps not surprising) to find that a mechanism whose purpose is to protect patients and serve their interests is fraught with corrupt people and practices. The self-interest of hospitals in maintaining their accreditation lies behind this newly disclosed pocket of unethical behavior.

The unethical behavior consists of rewriting minutes of meetings and other hospital records that are part of the quality assurance program.[3] The justification offered for the blatantly dishonest practice of altering minutes of meetings is that minutes are often jotted down in a hurried fashion and the need exists to make them "presentable" to inspectors who examine them. Some experts in quality assurance hold that no rewriting of minutes or records is proper, while others apparently allow minor corrections in grammar and some reorganization of written material.

Woodhull Medical and Mental Health Center in Brooklyn, New York, lost its accreditation after it was discovered that the hospital administration made changes in the minutes of the committee that reviewed surgical procedures in the hospital. The rewriting was designed to cover up failures in the quality assurance program and mistakes in medical care.

That was not an isolated incident. Evidence has been uncovered to reveal that the practice of rewriting minutes of meetings and altering other hospital records is widespread. There are consulting firms whose mission is to help hospitals rewrite their records at the last minute before pending inspections. Consultants who head these firms contend that they are simply "reformatting" existing material, presenting minutes of meetings in a manner consistent with the preferences of the JCAHO.

Yet it is clear that rewriting often goes beyond reformatting. An episode reported in the *New York Times* in March 1992 involved a second Brooklyn hospital, Kings County, in which the administration was worried about passing inspection. After it determined that the quality assurance program was lacking, a consulting firm named Quality Assurance Management Associates was asked to propose a solution. The consulting firm advised that the existing minutes of meetings would not pass inspection by JCAHO, and suggested that the hospital should augment the existing skimpy minutes by adding any conversations about cases that took place at other meetings, in the hallways of the hospital or the lunchroom. The firm's advice apparently rested on the judgment that there was no need to distinguish these other conversations from those that actually occurred at the committee meetings at which minutes were kept.[4]

At the time this episode occurred, the hospital had a new associate medical director for quality management, Dr. Robert M. Levine. Dr. Levine was an expert with years of experience in the field. He protested that these various proposals to rewrite the records amounted to fraud or, at the very least, a departure from honest documentation. What happened to Dr. Levine is one more instance in a long stream of examples of the fate of whistle blowers.

Dr. Levine objected to the hospital's plan to hire the consulting company, whose proposal, in his view, would result in the hospital's committing fraud. His strong objections dissuaded the hospital's executive director from hiring the company, and so the rewriting was to be done by personnel at the hospital. Both the original minutes and the rewritten version were to be given to inspectors from the commission, with changes to be dated, initialed, and signed off on by chairpersons of the departmental quality assurance committees.

When inspectors from JCAHO arrived at the hospital, they were given only the revised minutes from almost all of the departmental committees. These versions were not signed by the chairpersons. When he became aware of this, Dr. Levine handed one of the inspectors a note stating that they were being misled by the hospital. The next day, when Dr. Levine arrived at work, hospital guards prevented him from entering his office to remove his detailed hospital files. Dr. Levine was told that he was being dismissed from his position. As for the hospital, the JCAHO announced that it had failed the survey and was in danger of losing its accreditation. Dr. Levine, quite clearly a man of principle, was fired because his insistence on adhering to ethical principles was viewed as a threat to the interests of the hospital he worked for.

Depending on your degree of cynicism, this cautionary tale has one of two morals. The moral deriving from a position of unprincipled self-interest dictates that one should never be a whistle blower. The moral that stems from a realistic assessment of the conduct of professionals, government officials, and leaders in industry and finance, is: Don't expect people to be honest when they think they can gain more from being dishonest.

My own reaction to this lamentable state of affairs is one of dismay at the absence of mechanisms that can protect and improve the care of hospital patients. Among all the bureaucratic schemes in place—utilization review, peer review organizations, managed care plans, and quality assurance programs—only the last has the specific aim of protecting the interests of patients. Yet the existence of flourishing consulting firms that advertise: "We don't just tell you about your deficiencies— we *fix them!*"[5] attests to the way quality assurance has been distorted from a patient-centered activity to a mechanism for making sure that the hospital retains its accreditation.

In a world of increasing bureaucratization, no one is immune from criticism. The JCAHO, which accredits health care facilities, has itself been faulted for having needlessly technical rules, for constantly changing those rules, and for wide variation in the interpretations of those rules by its own inspectors. Even if those criticisms are valid, they cannot justify blatant fraud and dishonesty on the part of hospital administrators eager to get good marks by inspectors.

JCAHO mandates have recently moved a step beyond their traditional focus on medical procedures and hospital plants and operations. A recent addition to JCAHO regulatory standards requires hospitals to have an ethics committee or other mechanism (for example, an ethics consultation service) for addressing and resolving ethical conflicts. I have some doubt about whether a health care institution that is forced to establish an ethics committee or create another suitable mechanism for systematic review of ethical problems is likely to meet that challenge effectively in response to a regulatory requirement.

I am reminded of a meeting some years ago at which the existence and nature of hospital ethics committees were being discussed. One physician described what had occurred in a hospital where he was on the staff. The hospital's CEO, a prominent physician, had become aware that many hospitals were forming ethics committees and that it was the thing to do. Not wanting to have to answer "no" to the question of whether his hospital had an ethics committee, the CEO promptly formed such a committee, named himself as chair, and proceeded never to convene a meeting. The new JCAHO mandate could possibly result in such sham committees being formed. Alternatively, an equally troubling and perhaps more likely scenario is the formation of ethics committees by the hospital administration, structured in a manner that would further the administration's agenda, with barely a nod in the direction of respecting patients' rights and promoting their interests.

The early years of bioethics saw a primary focus on the physician–patient relationship. Identifying and articulating the rights of patients occupied the attention of many participants in the field. As bioethics matured, it was not uncommon to hear physicians at conferences defending the autonomy of patients and their right to be full participants in decision making. Now, less than three decades later, attention to autonomy and the rights of patients has waned. This trend does not signal a return to the old days of paternalism, in which physicians were in charge and patients were subordinate. Instead, the decline of interest in autonomy is more a function of the obsession with the costs of medical care.

So it is that the autonomy of patients and physicians alike is under fire. Granting autonomy to patients is seen as dangerous by those eager to set limits on costs. Patients and their families who ask that everything be done have to realize that they are not entitled to treatments that are medically futile. Then futility is defined so broadly or vaguely that it encompasses whatever physicians or hospital administrators claim is not medically appropriate or cost-effective.

The autonomy of physicians is under fire because of the perception that they have abused the health care system by doing too many diagnostic tests, overtreating patients, performing procedures that yield too little benefit, or ordering unnecessarily prolonged stays in the hospital. Superseding the professional autonomy of physicians to recommend what they believe is best for their individual patients, hospital administrators, risk managers, and bureaucrats at a distance have newfound authority to oversee and constrain physicians' decisions. All this is being done in a desperate attempt to control costs, an attempt that is bound to fail because the root causes of the high cost of health care in this country cannot be traced to the exercise of autonomy by patients or physicians.

Respect for autonomy is a fundamental ethical principle, though not a moral absolute. To grant unbridled autonomy to physicians to practice as they wish could endanger patients. A properly run system of quality assurance in hospitals is needed for weeding out incompetent or unscrupulous physicians and other substandard features of medical or nursing care. But when quality assurance programs become so enmeshed in shoring up the image of hospitals that they turn to fraud and other shady practices, everyone stands to be worse off: The hospital is in danger of losing its accreditation, patients are not protected by the very mechanism designed to ensure an adequate quality of care, and administrators who allow or promote dishonest practices are likely to suffer the consequences they deserve.

It is hard to see why anyone would favor a decline in the autonomy of patients after a reasonably successful struggle over many years to guarantee recognition and respect for patients' rights. There is one circumstance that could justify placing some limits on the autonomy of patients and physicians: a societal agreement to ensure a more just system for delivering health care. It is a serious mistake to think that reducing the overall costs of health care in the United States will have any effect on the injustices that now pervade the system. Controlling costs requires, as a first step, a concerted effort to achieve a single-payer system of reimbursement. That can be done even if the present modes of health care delivery are kept in place.

In contrast, rectifying injustices in access to health care requires some restructuring of the present modes of delivery. Striving to achieve greater justice does not require abandoning the traditional values of the physician–patient relationship or the more recently recognized autonomy of the patient. Although some people might be happy to see physicians' autonomy eroded, it is important to recognize that as the professional autonomy of doctors is weakened, so too is their ability to advocate vigorously for their patients' interests.

Notes

1. Susan M. Wolf et al., "Sources of Concern About the Patient Self-Determination Act," *New England Journal of Medicine*, vol. 325 (1991), pp. 1666–71.

2. Ibid.

3. Details about the episodes that follow are taken from an article written by Martin Gottlieb with Dean Baquet, "Questions of Ethics Trouble Hospitals Facing Inspection," *New York Times* (March 12, 1992), pp. A1, B4.

4. Ibid., p. B4.

5. Ibid.

Index